WHO Food Additives Series: 22

Toxicological evaluation of certain food additives

Toxicological evaluation of certain food additives

Prepared by

The 31st Meeting of the Joint FAO/WHO Expert Committee on Food Additives

Geneva 16-25 February 1987

Published on behalf of

The World Health Organization

The right of the
University of Cambridge
to print and sell
all manner of books
was granted by
Henry VIII in 1534.
The University has printed
and published continuously
since 1584.

CAMBRIDGE UNIVERSITY PRESS

Cambridge

New York   New Rochelle

Melbourne   Sydney

Published by the Press Syndicate of the University of Cambridge
The Pitt Building, Trumpington Street, Cambridge CB2 1RP
32 East 57th Street, New York, NY 10022, USA
10 Stamford Road, Oakleigh, Melbourne 3166, Australia

First published 1988
Reprinted 1989

Library of Congress cataloging in publication data available.

British Library cataloguing in publication data
Joint FAO/WHO Expert Committee on Food Additives,
Meeting (31st : 1987 : Geneva).
Toxicological evaluation of certain food additives and
contaminants.
1. Food additives — Toxicology   2. Food contamination
3. Food poisoning
I. Title   II. World Health Organization

ISBN 0 521 36928 2

The preparation of this document was supported by the International Programme on
Chemical Safety (IPCS), Geneve, Switzerland.

Transferred to digital printing 2003

CONTENTS

PREFACE

The monographs contained in this volume were prepared by the thirty-first Joint FAO/WHO Expert Committee on Food Additives (JECFA), which met in Geneva, Switzerland, 16-25 February 1987. These monographs summarize the safety data on selected food additives by the Committee. Generally, the compounds on which monographs were prepared are those on which substantial safety data exist. The data reviewed in these monographs form the basis for acceptable daily intakes (ADIs) established by the Committee.

The thirty-first report of JECFA has been published by the World Health Organization in the WHO Technical Report Series (No. 759). The participants in the meeting are listed in Annex 3 of the present publication and a summary of the conclusions of the Committee is included as Annex 4.

Specifications established by the thirty-first meeting of JECFA have been issued separately by FAO under the title *Specifications for the identity and purity of certain food additives*, FAO Food and Nutrition Paper, No. 38. These toxicology monographs should be read in conjunction with the specifications and the report.

Reports and other documents resulting from previous meetings of the Joint FAO/WHO Expert Committee on Food Additives are listed in Annex 1.

JECFA serves as a scientific advisory body to FAO, WHO, their Member States, and the Codex Alimentarius Commission, primarily through the Codex Committee on Food Additives, regarding the safety of food additives and contaminants in food. Committees accomplish this task by preparing reports of their meetings and publishing specifications and toxicological monographs, such as those contained in this volume, on substances that they have considered.

The toxicological monographs contained in this volume are based upon working papers that were prepared by temporary advisers in advance of the thirty-first JECFA meeting. A special acknowledgement is given to those who prepared these working papers: **Dr C.L.Galli**, Professor of Experimental Toxicology, University of Milan, Milan, Italy; **Dr S.I.Shibko**, Associate Director for Regulatory Evaluation, Center for Food Safety and Applied Nutrition, Food and Drug Administration, Washington, DC, USA; and **Dr Ronald Walker**, Professor of Biochemistry, University of Surrey, Guildford, Surrey, England.

Many proprietary unpublished reports are referenced. These were voluntarily submitted to the Committee by various producers of the food additives under review and in many cases these reports represent the only safety data available on these substances. The temporary advisers based the working papers they developed on all the data that were submitted, and all these studies were available to the Committee when it made its evaluations.

From 1972 to 1975 the toxicology monographs prepared by Joint FAO/WHO Expert Committees on Food Additives were published by WHO in the WHO Food Additives Series; after 1975 this series became available only in the form of unpublished WHO documents provided on request by the Organization. Beginning with the 1985 monographs, they are published by Cambridge University Press, which should ensure that these monographs are more widely known and available.

The preparation and editing of the monographs included in this volume have been made possible through the technical and financial contributions of the Participating Institutions of the International Programme on Chemical Safety (IPCS), which support the activities of JECFA. IPCS is a joint venture of the United Nations Environment Programme, the International Labour Organisation, and the World Health Organization, which is the executing agency. One of the main objectives of IPCS is to carry out and disseminate evaluations of the effects of chemicals on human health and the quality of the environment.

The designations employed and the presentation of the material in this publication do not imply the expression of any opinion whatsoever on the part of the organizations participating in the IPCS concerning the legal status of any country, territory, city, or area or its authorities, or concerning the delimitation of its frontiers or boundaries. The mention of specific companies or of certain manufacturers' products does not imply that they are endorsed or recommended by those organizations in preference to others of a similar nature that are not mentioned.

Any comments or new information on the biological or toxicological data on the compounds reported in this document should be addressed to: Joint WHO Secretary of the Joint FAO/WHO Expert Committee on Food Additives, International Programme on Chemical Safety, World Health Organization, Avenue Appia, 1211 Geneva 27, Switzerland.

ENZYME PREPARATIONS

ENZYME PREPARATIONS

Problems in evaluating the safety of enzymes in food processing were discussed at the fifteenth, eighteenth and twenty-ninth meetings of the Expert Committee, when principles relating to their evaluation were elaborated (Annex 1, references 26, 35, and 70). At its present meeting, the Committee reaffirmed those principles, which have been consolidated in Annex III of "Principles for the Safety Assessment of Food Additives and Contaminants in Food" (Annex 1, reference 76).

For the purpose of toxicological evaluation, the enzyme preparations under present consideration were grouped into the following classes:

- Class III - Enzymes derived from Aspergillus oryzae;
- Class IV - Enzymes derived from Aspergillus niger; and
- Class V   - Enzymes derived from Trichoderma reesei, Trichoderma harzianum, Penicillium funiculosum, Aspergillus alliaceus.

The guidelines established by JECFA for these classes of enzymes provide a basis for the toxicological studies required for their evaluation.

At the twenty-ninth meeting the Committee concluded that, when enzyme preparations from either class IV or class V are added directly to food but not subsequently removed, an acceptable daily intake should be established to ensure that levels of the enzyme preparations in food are safe. In order to evaluate the information received on the estimate of the amount of enzyme preparations used in the toxicological studies and levels of consumption resulting from their use in food, the Committee adopted the concept of enzyme total organic solids (TOS), which is defined as follows: % TOS = 100-(A+W+D), where A = % ash, W = % water, and D = % diluent and carrier (Ad hoc Enzyme Technical Committee, 1981; Pariza & Foster, 1983). This concept overcomes the problem that enzyme preparations of different activities and forms were used in the toxicological studies. It also takes into account that most of the organic solids in this fraction are not the enzyme per se.

In establishing acceptable daily intakes for the enzymes in classes IV and V, the Committee noted that the animal feeding studies were primarily of short-term duration. It, therefore, concluded that it would be appropriate to use a safety factor greater that the usual 100.

REFERENCES

Ad hoc Enzyme Technical Committee (1981). The 1978 enzyme survey, summarized data, National Academy of Sciences/National Research Council/Food and Nutrition Board, Committee on GRAS List Survey, Phase III, National Academy Press, Washington, D.C.

Pariza, M.W. & Foster, E.M. (1983). Determining the Safety of Enzymes used in Food Processing, J. Food Protection, 46: 453-468.

ENZYMES DERIVED FROM ASPERGILLUS ORYZAE

EXPLANATION

Enzymes from this source were considered at the fifteenth meeting of the Committee (Annex 1, reference 26), at which time a decision on the ADI was postponed because of concern that one of the known metabolites of A. oryzae is β-nitropropionic acid, which was suspected of carcinogenic potential. Later, at the eighteenth meeting of the Committee, a lipase derived from this organism was considered (Annex 1, reference 35). It was determined at that time that there was no information to substantiate the concern for the potential carcinogenicity of β-nitropropionic acid, and that analyses of foods have shown that the metabolite is present in very few foods and then only in minute amounts. The present Committee was also informed that A. oryzae varieties are used in certain parts of the world in the preparation of foods.

α-AMYLASE (E.C. 3.2.1)

BIOLOGICAL DATA

Biochemical aspects

No information available.

Toxicological studies
Acute toxicity

Animal	Route	LD_{50}	Reference
Mouse (Novo Strain)	Oral	> 20 g/kg b.w.	Novo, 1971a

Short-term studies
Rats

Three groups, each containing 5 male and 5 female SPF Wistar rats, were maintained for 3 weeks on diets containing 0, 0,5, or 5% of the enzyme preparation. Only minor differences were observed among the groups in body-weight change and food intake. At termination of the study, haematologic measurements, organ-weights analyses, and gross post mortem examinations showed no compound-related effects (Novo, 1971b).

In another study, two groups, each containing 10 male and 10 female ARS Sprague-Dawley rats, were fed diets containing 5 or 10% of the test enzyme (equivalent to 3.5 or 7 g enzyme/kg/b.w./day) for 90-94 days. A control group of 20 male and 20 female rats was maintained on the diet alone. No signs of toxicity were observed during the test period. Body-weight gain and food consumption were similar among animals in the test and control groups. Differential blood counts were within the normal range at weeks 4 and 8 in all groups. At the end of the study, haematologic parameters, organ-weight analyses, and gross and microscopic pathology showed no compound-related effects (Garvin <u>et al</u>., 1972a).

A similar study was performed with carbohydrases from <u>A. oryzae</u> (α-amylase and amyloglucosidase), prepared under different culture conditions. No compound-related effects were reported (Gavin <u>et al</u>., 1972b).

Long-term studies
No information available.

Observations in man

No information available.

COMMENTS

Short-term studies on α–amylase from __A. oryzae__ did not reveal any adverse effects. Based upon its lack of toxicity and the fact that __A. oryzae__ varieties are used in the preparation of foods, this enzyme was considered to be acceptable for use in food.

EVALUATION

Level causing no toxicological effects

Rat: 10% in the diet, equivalent to 7 g/kg b.w./day.

Estimate of acceptable daily intake

Acceptable for use in food when used according to good manufacturing procedures.

REFERENCES

Garvin, P.J., Ganote, C.E., Merubia, J., Delahany, E., Bowers, S., Varnado, A., Jordan, L., Hatley, G., DeSmet, C., & Porth, J. (1972a). Unpublished report from Travenol Laboratories, Inc., Morton Grove, IL, USA. Submitted to WHO by Gist-brocades NV, Delft, Holland.

Garvin, P.J., Ganote, C.E., Merubia, J., Delahany, E., Varnado, A., Jordan, L., Hatley, G., DeSmet, C., & Porth, J. (1972b). Carbohydrase from __A. oryzae__. Unpublished report from Travenol Laboratories, Inc., Morton Grove, IL, USA. Submitted to WHO by Gist-brocades NV, Delft, Holland.

Novo (1971a). Acute toxicity of fungamyl to mice. Unpublished report from Novo Industri A/S, Bagsvaerd, Denmark. Submitted to WHO by Novo Industri A/S, Bagsvaerd, Denmark.

Novo (1971b). Three week oral toxicity study of fungamyl in rats. Unpublished report BSi/BS from Novo Industri A/S, Bagsvaerd, Denmark. Submitted to WHO by Novo Industri A/S, Bagsvaerd, Denmark.

PROTEASES (E.C. 3.4.21.14; 3.4.23.6)

BIOLOGICAL DATA

Biochemical aspects

No information available.

Toxicological studies

Acute toxicity

No information available.

Short-term study

Rats

Two groups of 10 male and 10 female ARS Sprague-Dawley rats were fed diets containing 5 or 10% of the test enzyme preparation (equivalent to 3.5 or 7 g enzyme preparation/kg b.w./day) for 90 to 94 days. A control group of 20 male and 20 female rats were maintained on the diet alone. No signs of toxicity were observed during the test period. Body-weight gain and food consumption were similar in animals in the test and control groups. Differential blood counts were within the normal range at weeks 4 and 8 in all groups. At the end of the study serum clinical chemistry parameters, organ weight analyses, and gross and microscopic pathology showed no compound-related effects (Garvin <u>et al</u>, 1972).

Long-term studies

No information available.

Observations in man

No information available.

COMMENTS

A short-term study in rats on a protease preparation from <u>A. oryzae</u> did not reveal any adverse effects. Based on its lack of toxicity and the fact that <u>A. oryzae</u> varieties are used in the preparation of foods, this enzyme was considered to be acceptable for use in food.

EVALUATION

Level causing no toxicological effect

Rat: 10% in the diet, equivalent to 7 g/kg b.w./day.

Estimate of acceptable daily intake

Acceptable for use in food when used according to good manufacturing procedures.

REFERENCE

Garvin, P.J., Ganote, C.E., Merubia, J., Delahany, E., Bowers, S., Varnado, A., Jordan, L., Hatley, G., DeSmet, C., & Porth, J. (1972). Protease from <u>Aspergillus oryzae</u>. Unpublished report from Travenol Laboratories, Inc., Morton Grove, IL, USA. Submitted to WHO by Gist-brocades NV, Delft, Holland.

ENZYMES DERIVED FROM ASPERGILLUS NIGER

EXPLANATION

A. niger is a contaminant of food and was not considered in the same light as those organisms regarded as normal constituents of food. It is necessary to show that the strains used in enzyme preparations do not produce mycotoxins.

Microbial carbohydrases prepared from some varieties of A. niger were evaluated at the fifteenth meeting of the Committee, at which time a temporary ADI "not limited" was established (Annex 1, reference 26). A toxicological monograph was prepared (Annex 1, reference 27). An adequate 90-day study in rats was requested. Since the previous evaluation, additional data have become available on a number of carbohydrases, which are summarized and discussed in the following monograph. These enzymes were considered by the Committee to encompass the carbohydrases previously considered. The previously published monograph has been expanded and reproduced in its entirety below.

AMYLOGLUCOSIDASES (E.C. 3.2.1.3)

BIOLOGICAL DATA
Biochemical aspects
No information available.

Toxicological studies

Special studies on aflatoxin-related effects

Ducklings

Four groups of 5 ducklings received in their diet 0, 1, 5, or 10% enzyme preparation for 29 days. Growth, feed consumption, survival, behaviour, and mean liver weights were comparable, in all groups. No gross or histopathological lesions of the liver were seen (FDRL, 1963a).

Four groups of 5 ducklings received in their diet 0, 1, 5, or 10% enzyme preparation for 29 days. Growth, feed consumption, survival, behaviour, and development were comparable in all groups. No gross liver lesions were seen at autopsy and mean liver weights of treated animals were similar to those of controls. Histopathology of the livers was normal. No toxic elements were noted (FDRL, 1963b).

Acute toxicity[1]

Species	Route	LD_{50} (mg/kg b.w.)	Reference
Mouse	oral	> 3,200	Hunt & Garvin, 1963
		> 4,000	Hunt & Garvin, 1971
		> 3,200	Willard & Garvin, 1968
		> 4,000	Garvin et al., 1966
Rat	oral	10,000	Gray, 1960
		31,600	Kay & Calendra, 1962
		> 3,200	Willard & Garvin, 1968
		> 4,000	Garvin et al., 1966
		12,500 - 20,000	Kapiszka & Hartnage, 1978
Rabbit	oral	> 4,000	Garvin et al., 1966
Dog	oral	> 4,000	Garvin et al., 1966

[1] These data were obtained with several different commercial enzyme preparations.

Short-term studies

Rats

Three groups of 10 male rats received 0, 0.5, or 5% enzyme preparation in their diets for 30 days. No adverse effects related to treatment were observed regarding growth, appearance, behaviour, survival, food consumption, haematology, organ weights, or gross pathology (Garvin et al., 1966).

Two groups of 10 male and 10 female rats received either 0 or 5% enzyme preparation in their diets daily for 91 days. No differences from controls were observed regarding appearance, behaviour, survival, weight gain, haematology, organ weights, or gross pathology (Garvin & Merubia, 1959).

Two groups of 10 male and 10 female ARS Sprague-Dawley rats were fed diets containing 5 or 10% of the test enzyme preparation (equivalent to 3.5 or 7 g enzyme preparation/kg b.w./day) for 90 to 94 days. A control group of 20 male and 20 female rats were maintained on the diet alone. No signs of toxicity were observed during the test period. Body-weight gain and food consumption were similar between test and control groups. Differential blood counts were within the normal range at weeks 4 and 8 of the study in both test and control animals. At the end of the study serum clinical chemistry parameters, organ weight analyses, and gross and microscopic pathology showed no compound-related effects (Garvin et al, 1972).

Long-term studies

No information available.

Observations in man

No information available.

COMMENTS

Several short-term feeding studies in rats on amyloglucosidase preparations from A. niger have been performed. One study, in which the preparation was fed at up to 10% of the diet, was considered to be acceptable by current standards. No compound–related effects were observed in this study or in duckling tests that were performed to investigate potential aflatoxin–related effects.

The evaluations by the Committee of the carbohydrases and the protease from **A. niger** are summarized at the end of this section.

REFERENCES

FDRL (1963a). Unpublished report No. 84600e. Submitted to WHO by Miles Laboratories, Inc., Elkhart, IN, USA.

FDRL (1963b). Unpublished report No. 84600f. Submitted to WHO by Miles Laboratories, Inc., Elkhart, IN, USA.

Garvin, P.J. & Merubia, J. (1959). Unpublished report. Submitted to WHO by Baxter Laboratories, Inc.

Garvin, P.J., Willard, R., Merubia, J., Huszar, B., Chiu, E., & Gilbert, C. (1966). Unpublished report. Submitted to WHO by Baxter Laboratories, Inc.

Garvin, P.J., Ganote, C.E., Merubia, J., Delahany, E., Bowers, S., Varnado, A., Jordan, L., Hatley, G., DeSmet, C., & Porth, J. (1972). Unpublished report from Travenol Laboratories, Inc., Morton Grove, IL, USA. Submitted to WHO by Gist–brocades NV, Delft, Holland.

Gray, E.H. (1960). Unpublished report. Submitted to WHO by Miles Laboratories, Inc., Elkhart, IN, USA.

Hunt, R.F. & Garvin, P.J. (1963). Unpublished report. Submitted to WHO by Baxter Laboratories, Inc.

Hunt, R.F. & Garvin, P.J. (1971). Unpublished report. Submitted to WHO by Travenol Laboratories, Inc., Morton Grove, IL, USA.

Kapiszka, E.L. & Hartnage, R.E. (1978). The acute oral toxicity of Diazyme concentrate and Diazyme 325 in the rat. Unpublished report No. 16 from Miles Laboratories, Inc., Elkhart, IN, USA. Submitted to WHO by Miles Laboratories, Inc., Elkhart, IN, USA.

Kay, J.H. & Calendra, J.C. (1962). Unpublished report. Submitted to WHO by Miles Laboratories, Inc., Elkhart, IN, USA.

Willard, R. & Garvin, P.J. (1968). Unpublished report. Submitted to WHO by Travenol Laboratories, Inc., Morton Grove, IL, USA.

β-GLUCANASE (E.C. 3.2.1.6)

BIOLOGICAL DATA

Biochemical aspects ·

No information available.

Toxicological studies   (The TOS of the enzyme preparation used for toxicity studies was 49%).

Special studies on mutagenicity

The enzyme preparation was tested for mutagenic activity using 5 strains of <u>Salmonella typhimurium</u> (TA98, TA100, TA1535, TA1537, and TA1538 both with and without metabolic activation (S-9 fraction). The preparation was not mutagenic or toxic at concentrations up to 40 mg/ml (McConville, 1980).

A cytogenic bone marrow study was performed using adult male Chinese hamsters. Groups of adult male hamsters received up to 5000 mg/ kg b.w./day of the enzyme preparation for 5 consecutive days. Treatment did not result in an increased frequency of chromosomal aberrations in bone marrow (McGregor & Willins, 1981).

Acute toxicity

Species	Route	Sex	LD_{50} (ml/kg b.w.)	Reference
Mouse (NMRI)	oral	M & F	30	Novo, 1978a
Rat (Wistar)	oral	–	28.1	Novo, 1978b

Short-term studies

Rats

Three groups, each containing 5 male and 5 female Wistar/Mol SPF rats, were dosed orally by gavage once a day for 14 days with enzyme preparation at dose levels equivalent to 2.5, 5.0, or 10 ml/kg b.w. No clinical changes were observed. Body-weight gains of test and control animals were similar. At termination of the study, measurements of organ weights showed no compound-related effects (Novo, 1978c).

In another study, 4 groups, each containing 15 male and 15 female Wistar/Mol SPF rats, were dosed by gavage once a day for 90 days with enzyme preparation at dose levels equivalent to 0, 2.5, 5.0, or 10 ml/kg b.w. Deaths, primarily in the high-dose group, appeared to be related to injury during dosing. No clinical signs were observed in the other test animals. Male rats in the high-dose group showed decreased weight gain and marked decrease in food intake. Haematology studies showed increased platelet counts and decreased clotting times in the high-dose group at week 6, but this effect was not apparent at week 12. No other effects were reported. Clinical chemistry and urinalysis values at weeks 6 and 12 were within the normal range. At termination of the study, organ weight analysis showed a marked increase in relative weights of the spleen and testes of the males in the high-dose group. Gross and histopathological examination of the principal organs and tissues showed no compound-related effects (Perry et al., 1979).

Dogs

Three groups, each containing one male and one female beagle dog, received single doses of 5, 10, or 15 ml/kg b.w. of the enzyme preparation over a 4-day period. Following a 7-day observation period the dogs were sacrificed and subjected to macroscopic post-mortem examination. No compound-related effects were observed, with the exception of vomiting during the first 4 days of the study. In another study, dogs were administered consecutive doses of 15 ml/kg b.w./day for 9 days, and 10 ml/kg b.w./day for 5 days. No deaths occurred during the course of the study. The only clinical sign noted was excessive salivation and emesis shortly after dosing. Body weights, electrocardiograms, haematological parameters, blood serum chemistry, organ weights, gross pathology, and histopathology showed no compound-related effects (Osborne et al., 1978).

In another study, three groups, each containing 3 male and 3 female beagle dogs, were dosed with the enzyme preparation by gavage once a day, seven days a week, for 13 weeks, at dose levels equivalent to 2, 5, or 8 ml/kg b.w./day. Two dogs in the high-dose group died during the course of the study, which the authors concluded was due to respiratory distress as a result of foreign material in the lungs. Vomiting was reported after dosing in the high-dose group. Haematological parameters at weeks 6 and 12 were within normal limits, with the exception of a significant increase in WBC count, specifically in the group mean neutrophil counts, in the high-dose group. Clinical chemistry values were within the normal range at weeks 8 and 12, with the exception of slight increases in blood glucose and cholesterol in the high-dose group. Urinalysis showed no compound-related effects. At termination of the study, organ-weight analyses and gross and histopathological examination of the principal organs and tissues showed no compound-related effects (Greenough et al., 1980).

Long-term studies
No information available.

Observations in man
No information available.

COMMENTS
This enzyme preparation was not genotoxic in microbial or in mammalian test systems. Short-term studies in rats and dogs resulted in no observed compound-related effects at levels up to 5 ml/kg b.w./day of enzyme preparation.

The evaluations by the Committee of the carbohydrases and the protease from A. niger are summarized at the end of this section.

REFERENCES

Greenough, R.J., Brown, J.C., Brown, M.G., Cowie, J.R., Maule, W.J., & Atken, R. (1980). β-Glucanase 13 week oral toxicity study in dogs. Unpublished report No. 1630 from Inveresk Research International, Musselburgh, Scotland. Submitted to WHO by Novo Industri A/S, Bagsvaerd, Denmark.

McConville, M. (1980). Testing for mutagenic activity with S. typhimurium strain TA98, TA100, TA1535, TA1537, and TA1538 of fungal β-glucanase. Unpublished report No. 1751 from Inveresk Research International, Musselburgh, Scotland. Submitted to WHO by Novo Industri A/S, Bagsvaerd, Denmark.

MGregor, D.B. & Willins, M.J. (1981). Cytogenic study in Chinese hamsters of fungal β-glucanase. Unpublished report No. 2023 from Inveresk Research International, Musselburgh, Scotland. Submitted to WHO by Novo Industri A/S, Bagsvaerd, Denmark.

Novo (1978a). Acute oral toxicity of β-glucanase given to mice. Unpublished report No. 1978-06-30 RKH/PNi from Novo Industri A/S, Bagsvaerd, Denmark. Submitted to WHO by Novo Industri A/S, Bagsvaerd, Denmark.

Novo (1978b). Acute oral toxicity of β-glucanase given to rats. Unpublished report No. 1978-07-17 RKH/PNi from Novo Industri A/S, Bagsvaerd, Denmark. Submitted to WHO by Novo Industri A/S, Bagsvaerd, Denmark.

Novo (1978c). Oral toxicity of β-glucanase given daily to rats for 14 days. Unpublished report No. 1978-08-21 RKH/PNi from Novo Industri A/S, Bagsvaerd, Denmark. Submitted to WHO by Novo Industri A/S, Bagsvaerd, Denmark.

Osborne, B.E., Cockrill, J.B., Cowie, J.R., Maule, W., & Whitney, J.C. (1978). Beta-glucanase, dog acute and maximum tolerated dose study. Unpublished report No. 1208 from Inveresk Research International, Musselburgh, Scotland. Submitted to WHO by Novo Industri A/S, Bagsvaerd, Denmark.

Perry, C.J., Everett, D.J., Cowie, J.R., Maule, W.J. & Spencer, A. (1979). β-glucanase toxicity study in rats (oral administration by gavage for 90 days). Unpublished report No. 1310 from Inveresk Research International, Musselburgh, Scotland. Submitted to WHO by Novo Industri A/S, Bagsvaerd, Denmark.

HEMI-CELLULASE

BIOLOGICAL DATA

Biochemical aspects

No information available.

Toxicological studies

Special studies on mutagenicity

The enzyme preparation was tested for mutagenic activity using Salmonella typhimurium strains TA98, TA100, TA1535, and TA1537 both with and without metabolic activation (S-9 fraction). The test substance was not mutagenic or toxic at concentrations up to 5 mg/plate (Clausen & Kaufman, 1983).

In an in vitro cytogenetic test using CHO-K1 cells, both with and without metabolic activation (S-9 fraction), the enzyme preparation at test levels up to 2.5 mg (dry wt)/ml did not induce chromosomal aberrations (Skovbro, 1984).

Acute toxicity

No information available.

Short-term studies

Rats

Four groups, each containing 5 male and 5 female Wistar MOL/W rats, were dosed by gavage once a day for 90 days with the enzyme preparation at doses equivalent to 0, 100, 333, or 1000 mg/kg b.w./day. No significant clinical changes were observed. Body-weight gain and food intake were similar among test and control animals. Haematologic and clinical chemistry measurements at termination of the study were within normal ranges. Post-mortem examinations, measurements of organ weights, and histopathology showed no compound-related effects. Slight increases in kidney and adrenal weights in the mid-dose group were not associated with histopathological effects, and did not show a dose response (Kallesen, 1982).

Long-term studies

No information available.

Observations in man

No information available.

COMMENTS

This enzyme preparation was not genotoxic in microbial or in mammalian test systems. In a limited 90-day study in rats, no effects were observed at the highest dose administered (1 g/kg b.w./day). This enzyme preparation contained high levels of pectinase. The pectinase enzyme preparation summarized below may be identical to this hemicellulase preparation, which provides added assurance of the safety of this preparation.

The evaluations by the Committee of the carbohydrases and the protease from <u>A. niger</u> are summarized at the end of this section.

REFERENCES

Clausen, B. & Kaufman, U. (1983). Unpublished report from Obmutat Laboratiet. Submitted to WHO by Grinsted Products A/S, Brabrand, Denmark.

Kallesen, T. (1982). A 90-day toxicity study. Unpublished report No. 10023 from Scantox Biological Laboratory Ltd., Denmark. Submitted to WHO by Grinsted Products A/S, Brabrand, Denmark.

Skovbro, A. (1984). <u>In vitro</u> mammalian cytogenetic test (according to OECD Guideline No. 473). Unpublished report No. 10398 from Scantox Biological Laboratory Ltd., Denmark. Submitted to WHO by Grinsted Products A/S, Brabrand, Denmark.

PECTINASE (E.C. 3.1.1.11; 3.2.1.15; 4.2.2.10)

BIOLOGICAL DATA

Biochemical aspects

No information available.

Toxicological studies (The TOS of the commercial
preparation is approximately 5%).

Acute toxicity

Species	Route	LD_{50} (ml/kg b.w.)	Reference
Rat	oral	18.8–22.1	Porter & Hartnagel, 1979

Short-term studies

Rats

Two groups of 10 male and 10 female ARS Sprague–Dawley rats
were fed diets containing 5 or 10% of the test enzyme preparation
(equivalent to 3.5 or 7 g of the enzyme preparation/kg b.w./day), for
90 to 94 days. A control group of 20 male and 20 female rats was
maintained on the diet alone. No signs of toxicity were observed
during the test period. Body-weight gain and food consumption were
similar among test and control groups. Differential blood counts at
weeks 4 and 8 of the study were within the normal range in test and
control animals. At the end of the study serum clinical chemistry
analyses, organ weight analyses, and gross and microscopic pathology
showed no compound-related effects (Garvin et al., 1972).

Long-term studies

No information available.

Observations in man

No information available.

COMMENTS

In a short-term study in rats, no adverse effects were observed at dietary levels of the enzyme preparation up to the equivalent of 7 mg/kg b.w./day. This enzyme preparation may be identical to the hemi-cellulase preparation summarized above. The hemi-cellulase enzyme preparation summarized above also contained high levels of pectinase, which provides added assurance of the safety of this preparation.

REFERENCES

Garvin, P.J., Ganote, C.E., Merubia, J., Delahany, E., Bowers, S., Varnado, A., Jordan, L., Hatley, G., DeSmet, C., & Porth, J. (1972). Carbohydrase from __Aspergillus niger__ (pectinase, cellulase and lactase). Unpublished report from Travenol Laboratories, Inc., Morton Grove, IL, USA. Submitted to WHO by Gist-brocades NV, Delft, Holland.

Porter, M.C. & Hartnagel R.E. (1979). The acute oral toxicity of a new pectinase product in the rat. Unpublished report No. 11 from Miles Laboratories, Inc., Elkhart, IN, USA. Submitted to WHO by Enzyme Technical Association, Washington, DC, USA.

PROTEASE

No information available.

GENERAL COMMENTS ON ENZYMES FROM A. NIGER

Aspergillus niger is a contaminant of food. Although there may be posible strain differences in A. niger, and different cultural conditions might be used to prepare the various enzymes, the available toxicity data, which consist primarily of short-term feeding studies in rats and some studies in dogs, show that all the enzyme preparations tested were of a very low order of toxicity. The enzyme preparations tested were non-mutagenic in bacterial and mammalian cell systems. Studies on some strains of A. niger used to prepare carbohydrases showed no aflatoxin or related substance production. These studies provide the basis for evaluating the safety of enzyme preparations derived from A. niger. It was also noted that the enzyme preparations tested exhibit a number of enzyme activities, in addition to the major enzyme activity. Thus, there may be considerable overlap of the enzyme activities of the different enzyme preparations so that safety data from each preparation provides additional assurance of safety for the whole group of enzymes.

Since the enzyme preparations tested were of different activities and forms, and most of the organic materials in the preparations are not the enzyme per se, the numerical ADI is expressed in terms of total organic solids (TOS) (see introduction to enzyme preparations section).

EVALUATION
Level causing no toxicological effect
All enzyme preparations tested showed no-observed-effect levels greater than 100 mg TOS/kg b.w./day in 90-day studies in rats.

Estimate of acceptable daily intake
0-1 mg TOS/kg b.w. for each of the enzyme preparations.

EXPLANATION

This enzyme preparation has not been evaluated previously by the Joint FAO/WHO Expert Committee on Food Additives. The preparation used in the toxicological studies were obtained by spray drying the enzyme preparation; the preparation contained 80 - 95% TOS.

BIOLOGICAL DATA

Biochemical aspects

No information available.

Toxicological studies

Special studies on mutagenicity

The enzyme preparation was tested for mutagenic activity using Salmonella typhimurium strains TA98, TA100, TA1535, and TA1537 and Escherichia coli WP 2uvra pmK 101 (CM891), both with and without metabolic activation (S-9 fraction). No dose-related increases in revertants were obtained at test levels up to 10 mg/ml of the incubation mixture (Pedersen, 1984).

A cytogenic bone marrow study was performed using adult male Chinese hamsters. Groups of adult male hamsters received up to 5000 mg/kg/day of the enzyme preparation for 5 consecutive days. Treatment did not result in an increased frequency of chromosomal aberrations in bone marrow (McGregor & Holmstrom, 1981).

The enzyme preparation was not mutagenic in mouse lymphoma L5178Y cells at concentrations up to 3.5 mg/ml, with or without metabolic activation (McGregor & Riach, 1984).

Special study on reproduction and teratogenicity

Rats

Groups of 8 pregnant Sprague-Dawley CD rats were used to establish the maternal embryonic maximum tolerated dose. The test animals were dosed daily from day 6 through day 16 of gestation and killed on day 20 of gestation; the number of corpora lutea graviditatis in each ovary and the number of and location of all implantations in the uterus were recorded. The maternal embryonic tolerated dose was established at 5 g/k.g. b.w. For the teratology study, four groups each containing 24 pregnant Sprague-Dawley rats, were dosed daily from day 6 through day 16 (day 17 for controls) with 0, 1, 3, or 8 g/kg b.w. of the enzyme preparation. The rats were killed on day 20 of gestation. Maternal weight gain was reduced at all dose levels of the test compound, and this was accompanied by reduced food consumption in the high-dose group. There were slight reductions in mean litter fetal weight and in mean litter placental weight, which were dose-related. There were no significant dose-related trends in pregnancy data (number of corpora lutea, implantations, resorptions, or live fetuses) or in skeletal abnormalities. However, both the 3 and 8 g/kg b.w. groups showed slight increases in the incidence of hydronephrosis, and the incidence of hydroureter was marginally increased in a dose-related manner. The incidences of these effects, although not significantly different from those in concurrent controls, exceeded the usual background in historical controls (Hazelden & Maddock, 1982).

Acute toxicity

Species	Route	LD_{50} (g/kg b.w.)	Reference
Mouse (Novo)	oral	> 20	Novo, 1975
Rat (Mollegard)	oral	> 10	Novo, 1982

Short-term studies

Rats

Four groups of 15 male and 15 female Charles River CD rats, 4 weeks of age, were maintained for 13 weeks on diet containing 0, 500, 1,500, or 5,000 mg/kg b.w./day of the enzyme preparation. No compound-related deaths were reported. Food intake and body-weight gain were similar in test and control groups, with a tendency for increased weight gain in females in the high-dose group during the last weeks of the test. Opthalmoscopic examinations at weeks 0 and 12 showed no abnormalities. Haematologic, clinical chemistry, and standard urinalysis values were within normal ranges. However, urinary alkaline phosphatase levels were increased in male rats in the two high-dose level groups at week 12 of the study. Liver cytochrome P-450 measurements showed no evidence of enzyme induction. At termination of the study, no compound-related changes were observed after organ weight analysis and gross and microscopic examination of the principal organs and tissues, with the exception of a slight increase in the incidence of inflammatory reactions in the kidney cortex in the high-dose groups (Warwick et al., 1976).

Dogs

Four groups of 3 male and 3 female dogs were dosed with the enzyme preparation by gavage once a day 7 days a week for 13 weeks at dose levels equivalent to 0, 300, 1000, or 3000 mg/kg b.w./day. Vomiting was reported in the high-dose group. No other clinical signs were observed. Decreased weight gain was observed in female dogs in the two high-dose groups. Although food consumption was also decreased in these groups, the reduced body weight in the high-dose group was greater than expected from the decreased food intake. Haematologic, clinical chemistry, and urinalysis values at weeks 6 and 12 showed no compound-related effects, with the exception of increased urinary alkaline phosphatase at week 12 of the study. At termination, gross necropsy, organ weight analysis, and microscopic examination of the principal organs and tissues showed no treatment-related effects. Liver cytochrome P-450 measurements showed no evidence of enzyme induction (Edwards et al., 1976).

Long-term studies

No information available.

Observations in man

No information available.

COMMENTS

The beta-glucanase preparation was not mutagenic in bacterial or in mammalian systems. The preparation caused no adverse effects in a reproduction study in rats at levels up to 5%, and it was not teratogenic in a rat study at doses up to 1 g/kg b.w./day. Short-term studies showed no adverse effects at 3 g/kg b.w./day in dogs or at 2 g/kg b.w./day in rats. Based on the available information, the Committee established a temporary ADI for this enzyme preparation.

Because this enzyme is derived from a microorganism that is neither a normal constituent of food nor a common contaminant in food, in accordance with Annex III of "Principles for the Safety Assessment of Food Additives and Contaminants in Food" (Annex 1, reference 76), this preparation requires the submission of results of a long-term study in a rodent species as well as specifications to show that the organism does not produce antibiotics and is non-pathogenic to man.

EVALUATION

Level causing no toxicological effect

Rat: 1000 mg/kg b.w./day (teratogenicity study).

Estimate of temporary acceptable daily intake

0-0.5 mg TOS/kg b.w.

Further work or information

Required (by 1992)

1. Long-term feeding study in a rodent species.

2. Additional information to show that this organism does not produce antibiotics and is non-pathogenic to man.

REFERENCES

Edwards, D.B., Osborne, B.E., Kinch, D.A., & Dent, N.J. (1976). Mutanase toxicity study in beagle dogs (oral administration for 13 weeks). Unpublished report No. 433 from Inveresk Research International, Musselburgh, Scotland. Submitted to WHO by Novo Industri A/S, Bagsvaerd, Denmark.

Hazelden, K. & Maddock, S. (1982). Teratogenicity study in rats. Unpublished report No. 2253 from Inveresk Research International, Musselburgh, Scotland. Submitted to WHO by Novo Industri A/S, Bagsvaerd, Denmark.

McGregor, D.B. & Holmstrom, L.M. (1981). Cytogenetic study in Chinese Hamsters of SP 116. Unpublished report No. 2208 from Inveresk Research International, Musselburgh, Scotland. Submitted to WHO by Novo Industri A/S, Bagsvaerd, Denmark.

McGregor, D.B. & Riach, C.G. (1984). SP 116 batch 1531, mouse lymphoma mutation assay. Unpublished report No. 2894 from Inveresk Research International, Musselburgh, Scotland. Submitted to WHO by Novo Industri A/S, Bagsvaerd, Denmark.

Novo (1975). Acute oral toxicity of mutanase to mice. Unpublished report No. F-751886.1 from Novo Industri A/S, Bagsvaerd, Denmark. Submitted to WHO by Novo Industri A/S, Bagsvaerd, Denmark.

Novo (1982). Acute toxicity of SP 116 (Batch PPM 1216) given once orally to rats. Unpublished report No. 5181 from Novo Industri A/S, Bagsvaerd, Denmark. Submitted to WHO by Novo Industri A/S, Bagsvaerd, Denmark.

Pedersen, P.B. (1984). Glucanex (batch No. PPM 1531), testing for mutagenic activity with <u>Salmonella typhimurium</u> and <u>Escherichia coli</u> WP2uvrA (pK M 101) in liquid culture assay. Unpublished report No. E.0184 from Novo Industri A/S, Bagsvaerd, Denmark. Submitted to WHO by Novo Industri A/S, Bagsvaerd, Denmark.

Warwick, M.H., Osborne, B.E., Collings, A.J., Kinch, D.A., & Dent, N.J. (1976). Mutanase toxicity study in rats (oral administration for 13 weeks). Unpublished report No. 461 from Inveresk Research International, Musselburgh, Scotland. Submitted to WHO by Novo Industri A/S, Bagsvaerd, Denmark.

CELLULASE FROM TRICHODERMA REESEI

EXPLANATION

This substance has not been evaluated previously by the Joint FAO/WHO Expert Committee on Food Additives. Cellulase is produced extracellularly by T. reesei (QM6a), a mutant of T. viride. The enzyme preparation is characterized by two activities, exo-cellobiohydrolase (E.C. 3.2.1.1) and 1,4-endoglucanase (E.C. 3.2.1.4).

Tests have been performed to show that the strain of T. reesei used for the production of this enzyme preparation does not produce any antibiotics, and it can be regarded as non-pathogenic.

The enzyme preparation used in the toxicological studies was a spray dried product derived from a cruder preparation than the commercial product. The TOS of the tested product was 31%.

The available safety data have been summarized by Hjortkjer et al., 1986.

BIOLOGICAL DATA

Biochemical aspects

No information available.

Toxicological studies

Special studies on mutagenicity

The mutagenicity of the enzyme preparation was tested using Salmonella typhimurium strains TA98, TA100, TA1535, TA1537 and TA1538, both with and without metabolic activation by S-9 fraction from rat

liver.   No mutagenic effects were observed at concentrations up to
10,0 mg/plate (Crichton & McGregor, 1977).

A cytogenetic bone marrow study was performed using adult
male Chinese hamsters.   Groups of adult male hamsters received up to
5000 mg/kg b.w./day of the cellulase preparation for 5 consecutive
days.   The cellulase preparation did not produce an increase in
chromosome aberrations in bone marrow cells (McGregor, 1979).

In a dominant lethal assay, adult male CD-1 mice were dosed
orally by gavage for 5 consecutive days at dose levels up to 5000 mg/kg
b.w./day.   No treatment-related effects were observed on implantation
or fetal deaths (Cuthbert et al., 1980).

Special study on reproduction

Rats

Four groups, each containing 20 male and 20 female 6-week
old CD rats, were maintained on diets containing 0, 1, 2, or 5% of the
enzyme preparation.   The test diet was fed for 10 weeks prior to
breeding and throughout mating, gestation, and lactation.   Weanlings
were maintained on the same diet as parents, until autopsied at 28 days
of age.

Parameters evaluated included body weights, feed consumption
of F_0 and F_1 animals, and reproduction parameters (fertility index,
gestation index, live birth index, litter size, viability index, and
lactation index).   Gross necropsies were conducted on all F_0 and F_1
animals, except pups dying before day 12 of lactation.   Organ weights
from 10 male and 10 female F_1 animals from each treatment group were
measured, and the principal tissues and organs from a similar number of
F_1 animals from the high-dose and control groups were microscopically
examined.   Clinical chemistry, haematology, and urinalysis were not
performed.   Compound-related mortality was not reported in the F_0
generation.   There were no treatment-related clinical signs.   Body
weights of males in the high-dose group were lower than those of
controls, which was associated with decreased food intake.   There were
no treatment-related effects on reproductive parameters.   In the F_1
generation, there was a trend to increased mean body weights during the

early period of lactation, but this effect was not significant toward
the end of the treatment period. No significant treatment-related
effects on absolute or relative organ weights in F_1 males and females
were observed, and no treatment-related gross or microscopic adverse
effects were reported (Hazelden et al., 1982).

Special study on teratogenicity

Three groups of 6 pregnant CD rats each were dosed by gavage
with 700, 2400, or 7000 mg/kg b.w./day of the enzyme preparation from
days 6 to 16 of gestation. The rats were killed on day 20 of gestation.
Reduced body-weight gain was observed in the high-dose group, which was
associated with decreased food consumption during the dosing period.
No compound-related differences were observed in placental weights,
number of corpora lutea, implantations, resorptions, or live fetuses.
The small number of visceral and skeletal abnormalities showed no
treatment associations or trends (Hazelden & Everett, 1980).

Acute toxicity

Species	Route	LD_{50} (g/kg b.w.)	Reference
Mouse (NMRI)	oral	16	Modeweg–Hansen, 1978a
Rat (Wistar)	oral	8	Modeweg–Hansen, 1978b
Dog	oral	5	Osborne & Chambers, 1977

Short-term studies

Rats

Four groups, each containing 15 male and 15 female CD rats
4 weeks of age, were maintained for 13 weeks on diets containing 0, 1,
2, or 5% of the enzyme preparation. Abnormal grooming in the high-dose
groups was observed during the first 8 weeks of the study. Decreased
weight gain from weeks 4 to 10 was observed in the high-dose group.
Blood urea values were elevated in treated animals at weeks 4 and 13,

but a consistent dose-response effect was not observed. There were no other compound-related effects on clinical chemistry or haematological measurements. No compound-related deaths were reported. At autopsy, male rats in the high-dose group had a significant increase in organ-to-body-weight ratios for the liver, prostate, and kidney when compared to control values, and females in the high-dose group had significantly higher spleen-to-body-weight ratios than controls. No compound-related gross or microscopic changes were observed, except in the case of the kidney of some rats in the high-dose group, where there was a small increase in the size of proteinaceous globules in the epithelial cells lining the renal convulated tubules (Ben-Dyke et al., 1977).

Dogs

Four groups, each containing 3 male and 3 female dogs, were dosed by gavage once a day, seven days a week, for 13 weeks with the enzyme preparation (30% dispersion in water) at doses equivalent to 0, 750, 1500, or 3000 mg/kg b.w./day. Vomiting was reported after dosing in the high-dose groups during the first 3 to 4 weeks of the study, and diarrhea was observed in these groups during the first two weeks of the study. There were no significant differences in body weight and food consumption between dosed and control animals during the course of the study. Haematology, clinical chemistry, and urinalysis at weeks 6 and 12 of the study showed no treatment-related effects. Ophthalmoscopic examinations at weeks 6 and 12 showed no treatment-related effects. At termination of the study, gross necropsies, organ weight analyses, and microscopic examinations of the principal organs and tissues showed no treatment-related effects (Osborne et al., 1977).

Long-term studies

No information available.

Observations in man

No information available.

COMMENTS

This cellulase preparation from T. reesei was not mutagenic in bacterial or in mammalian systems. The preparation caused no adverse effects in a reproduction study in rats at levels up to 5% in the diet, and it was not teratogenic in a rat study at doses up to 7 g/kg b.w./day. Short-term studies are available in dogs and rats, the no-adverse-effect levels being 3 g/kg b.w./day in dogs and 2 g/kg b.w./day in rats. The Committee was also informed that tests have been performed to show that the strain of T. reesei used for the production of this enzyme does not produce any antibiotics and it is not known to be a human pathogen. Based on the available information, the Committee established a temporary ADI for this enzyme preparation.

Because this enzyme is derived from a microorganism that is neither a normal constituent of food nor a common contaminant in food, in accordance with Annex III of "Principles for the Safety Assessment of Food Additives and Contaminants in Food" (Annex 1, reference 76), this preparation requires the submission of the results of a long-term study in a rodent species.

EVALUATION

Level causing no toxicological effect

Rat: 20,000 ppm in the diet, equivalent to 2000 mg/kg b.w./day (600 mg TOS/kg b.w./day).

Estimate of temporary acceptable daily intake for man

0-0.3 mg TOS/kg b.w.

Further work or information

Required (by 1992)

Long-term feeding study in a rodent species.

REFERENCES

Ben-Dyke, R., Strachen, E., Kiss, I., & Finn, J.P. (1977). Cellulase: toxicity in dietary administration to rats for thirteen weeks. Unpublished report No. 77/NTL-33/382 from Life Science Research, Stock, England. Submitted to WHO by Novo Industri A/S, Bagsvaerd, Denmark.

Crichton, C., & McGregor, D.B. (1977). Testing for mutagenic activity in cellulase® (SP-122). Unpublished IRI project No. 40849 from Inveresk Research International, Edinburgh, Scotland. Submitted to WHO by Novo Industri A/S, Bagsvaerd, Denmark.

Cuthbert, J.A., McGregor, D.B., & Willins, M.J. (1980). Dominant lethal study in mice of acid cellulase. Unpublished IRI project No. 702021, report No. 1699, from Inveresk Research International, Edinburgh, Scotland. Submitted to WHO by Novo Industri A/S, Bagsvaerd, Denmark.

Hazelden, K.P. & Everett, A. (1980). Teratogenicity testing in rats of acid cellulase. Unpublished IRI project No. 702016 from Inveresk Research International, Edinburgh, Scotland. Submitted to WHO by Novo Industri A/S, Bagsvaerd, Denmark.

Hazelden, K.P., Maddock, S.M., & Rushton, A.K.A. (1982). Acid cellulose dietary toxicity study in rats with in utero exposure. Unpublished IRI project No. 704851, report No. 2350, from Inveresk Research International, Musselburgh, Scotland. Submitted to WHO by Novo Industri A/S, Bagsvaerd, Denmark.

Hjortkjer, R.K., Bille-Hansen, V., Hazelden, K.P., McConville, M., McGregor, D.B., Cuthbert, J.A., Greenough, R.J., Chapman, E., Gardner, J.R., & Ashby, R. (1966). Safety of celluclast®, an acid cellulase derived from Trichoderma reesei. Fd. Chem. Tox., 24, 53-63.

McGregor, D.B. (1979). Cytogenetic study in chinese hamsters of acid cellulase. Unpublished IRI project No. 702042, report No. 1585, from Inveresk Research International, Edinburgh, Scotland. Submitted to WHO by Novo Industri A/S, Bagsvaerd, Denmark.

Modeweg-Hansen, L. (1978a). Acute oral toxicity of cellulase SP-122 to rats. Unpublished report from Novo Industri A/S. Submitted to WHO by Novo Industri, Bagsvaerd, Denmark.

Modeweg-Hansen, L. (1978b). Acute oral toxicity of cellulase SP-122 to mice. Unpublished report from Novo Industri A/S. Submitted to WHO by Novo Industri, Bagsvaerd, Denmark.

Osborne, B.E. & Chambers, P.R. (1977). Cellulase SP-122, acute oral toxicity study in dogs. Unpublished IRI project No. 408473 from Inveresk Research International, Edinburgh, Scotland. Submitted to WHO by Novo Industri A/S, Bagsvaerd, Denmark.

Osborne, B.E., Rushton, A.K.A., & Dent, N.J. (1977). Cellulase SP-122, toxicity study in beagle dogs (oral administration for 13 weeks). Unpublished IRI project No. 408489, report No. 919, from Inveresk Research International, Edinburgh, Scotland. Submitted to WHO by Novo Industri A/S, Bagsvaerd, Denmark.

FLAVOURING AGENT

SMOKE FLAVOURINGS

EXPLANATION

Smoke condensates and liquid smoke were considered at the nineteenth meeting of the Joint FAO/WHO Committee on Food Additives (Annex 1, reference 38). Inadequate information was available at that time for an evaluation.

The present Committee reviewed both the specifications and safety data for this group of products. It noted that smoke flavourings were complex mixtures of varying composition, primarily prepared by the condensation of smoke generated by pyrolysis of certain hardwoods in the absence of or in the presence of a limited amount of air. Woods commonly used for the preparation of smoke flavourings include oak (Quercus spp.), hickory (Carya spp.), beech (Fagus spp.), alder (Alnus spp.), and maple (Acer spp.). No toxicity data are available on pyroligneous acid preparations which are condensates derived from the pyrolysis of wood in the absence of air. The smoke flavourings reviewed in this monograph are derived from the smoke condensate of wood burned in a limited amount of air.

The initial smoke condensate separates into an aqueous phase and a tarry phase. The smoke condensate may be separated into fractions by physical separation techniques or solvent extraction. These fractions may be further purified, if necessary, to remove hazardous constituents known to be present in smoke. Smoke flavourings include smoke condensates, fractions thereof, and mixtures of such fractions.

SMOKE FLAVOURINGS (AQUEOUS PHASE)

BIOLOGICAL DATA

Biochemical aspects

No information available.

Toxicological studies

Special study on carcinogenicity

An aqueous wood smoke flavouring, which previously had been shown to be inactive in a <u>Salmonella typhimurium</u> mutation assay but positive in an assay utilizing TK6 human lymphoblasts, was tested in the mouse lung adenoma assay. In this study, newborn Swiss Webster mice were injected with the smoke flavouring on days 1, 8, and 15, and then maintained until 26 weeks of age. The total dose of test substance administered ranged from 17.5 µl to 31.5 µl. The test substance did not induce lung tumours or tumours at other sites. Toxicity was observed at 15 weeks in some of the test animals, which included hyperplastic kidneys and abnormalities in the colon and rectum (Braun <u>et al</u>, 1986).

Special studies on mutagenicity

Smoke flavouring preparations were tested using <u>Salmonella typhimurium</u> strains TA98, TA100, TA1535, TA1537, and TA1538 and in <u>E. coli</u> WP2 (UVRA), both with and without metabolic activation, at doses ranging from 3 to 10,000 µg/plate. Toxicity was observed at the highest dose tested without activation. The preparations were not mutagenic in these tests (Mortelmans & Eckford, 1980; Rattech, 1981).

An aqueous wood smoke flavouring which was not mutagenic in the <u>Salmonella typhimurium</u> forward mutation assay (using strain TM677), at concentrations up to 18 µg/ml (of dissolved solids), induced a significant increase in mutation frequency of TK6 human lymphoblasts (Braun <u>et al.</u>, 1986).

Acute toxicity

No information available.

Short-term studies

Rats

Four groups, each containing 10 male and 10 female rats, were maintained for 90 days on diets containing 0, 0.3, 2.5, or 20% of a liquid smoke preparation (maple or hickory). All diets were adjusted to contain 20% added water. Slight growth reduction was observed at the high-dose level, but this was associated with decreased feed intake. Haematologic and urine analyses at the end of the study showed no abnormalities in any of the test groups. At autopsy, organ-weight analyses of the principal organs and tissues showed no significant changes in absolute organ weights. Gross and microscopic examinations of these tissues showed no compound-related effects (WARF, 1961).

Three groups of 25 male and 25 female rats were maintained for 90 days on diets containing 0, 0.25, or 2.0% of one liquid smoke flavouring or 2.0% of another smoke flavouring preparation. Growth and food intake were similar among test and control groups. Haematologic values at week 6 and at termination of the study were within normal limits in all groups. At autopsy, organ-weight analyses and gross pathology and histopathology of the principal organs were carried out. No compound-related effects were observed, with the exception of minor degenerative changes in the liver and kidneys in one of the 2.0% dose groups and slight bone marrow hypoplasia in both 2.0% dose groups (WRC, 1963; FCT, 1965).

Analytical studies on the preparation fed at 0.25 and 2.0% of the diet showed the absence of 3,4-benzo(a)pyrene, but the presence of 5 non-carcinogenic polycyclic aromatic hydrocarbons, namely benzo(a) anthracene, carbazole, chrysene, pyrene, and fluouranthene at levels of 0.012, 0.20, 0.011, 0.013, and 0.032 ppm of the liquid smoke preparation, respectively. The other preparation contained formic, propionic, vanillic, and siringic acids, dimethylphenol, methyl glyoxal, furfural, acetaldehyde, acetone, ethanol, and benzopyrenes (< 1 ppb

benzo(a)pyrene, 6 ppb pyrene, 3 ppb flouranthene, and 18 ppb phenan-
threne (Lijinsky & Shubik, 1965).

A series of studies was carried out with two liquid smoke
flavourings. The concentrations of total acids, phenols, and carbonyl
compounds in these preparations are outlined in Table 1.

Table 1. Constituents of liquid smoke 1 (LS 1) and liquid smoke 2 (LS 2)

Constituent	LS 1	LS 2
Total acids (acetic acid)	1.8 g/l	2.0-2.8 g/l
Total carboxylic compounds	2.0 g/l	3.0-3.6 g/l
Phenols	0.018 g/l	0.9-1.3 g/l
PAHs (3,4-benzo(a)pyrene)	--	0.2 ppb

Five groups, each containing 6 male and 6 female Colworth
Wistar rats, were fed diets containing 1.25, 2.5, 5, 10, or 20% LS 2 or
20% LS 1 for 28 days. Another group of 12 male and 12 female rats was
fed control diets. Body-weight gains of both male and female rats fed
20% of either preparation, or 5 or 10% LS 2, were less than those of
controls. Plasma analysis at termination of the study showed sporadic
variation in a number of constituents, but these variations were not
dose-related. However, decreased plasma alkaline phosphatase was
observed at all dose levels in male rats and female rats in the
high-dose groups showed elevated thyroxine, (T_4) and triiodothyronine
(T_3) levels. Haematologic changes at the high-dose level included
decreased platelet counts in male rats administered LS 1 and in
haemoglobin and red cell counts in male rats fed LS 2. At autopsy,
organ-weight analysis showed increased relative liver and kidney
weights in high-dose male rats fed LS 2 and increased thyroid weight in
high-dose female rats fed LS 2. Histopathological examination of
tissues and organs showed no compound-related effects, with the
exception of severe corticomedullary nephrocalcinosis in females fed
20% LS 2 (Parish, 1986a).

In a paired feeding study, groups of 6 male and 6 female Colworth Wistar rats were fed diets containing 2 or 20% LS 1 or LS 2. A control group of 12 male and 12 female rats was fed control diet ad libitum. Minor differences in weight gain were observed between test animals and paired controls. At autopsy, serum chemistry values of test animals and paired controls were similar. Organ-weight analysis showed significantly, increased liver and kidney weights in rats fed liquid smoke when compared to paired controls. Macroscopic evaluation of organs and microscopic examination of the livers and kidneys showed no compound-related effects (Parish, 1986b).

Three groups of 20 male and 20 female Carworth Wistar rats were maintained on diets containing LS 2 for 13 weeks. The levels of liquid smoke in the diet were adjusted biweekly, so that the intake in mg/kg b.w. was constant. The levels of liquid smoke in the diet were 0.5-1.0, 1.5-3.0, and 3.0-6.0%, which were equivalent to daily intakes of 0.6, 1.9, and 3.8 g/kg b.w./day for males and 0.7, 2.2, and 4.4 g/kg b.w./day for females, respectively. Additional groups at the high-dose level and the control group were maintained for another 13 weeks on control diets. No animals died during the course of the study. No significant differences in body weights were observed at weeks 13 or 26, although there were significant reductions in body-weight gain during weeks 4-8 in both male and female rats in the high-dose group. At termination of the study, serum analysis showed a number of changes in both sexes in the high-dose group. Haematologic changes were also observed in this group, which included increased haemoglobin and packed RBC volume in male rats and decreased WBC counts and lymphocytes in females. Organ-weight analyses showed significant increases in relative liver and kidney weights of male and female rats in the high-dose group. Relative liver weights of female rats were also increased in the 1% group. Histopathology of the tissues and organs at week 13 of the study showed no treatment-related changes. All the changes observed in rats at 13 weeks were no longer apparent after the 13-week recovery period (Parish, 1986c).

In a study designed to investigate possible effects of smoke flavouring on the thyroid, groups of 3 male and 3 female rats received diets containing from 0.625 to 20% LS 1 for 28 days. Another group was administered 0.025% 2-thiouracil in the diet, and another group of 6 male and 6 female rats was fed control diet. Body-weight gains of female rats fed 20% liquid smoke were significantly lower than those of control animals. Food intake of rats in the 2-thiouracil group was significantly depressed. At termination, plasma studies, including determination of thyroid hormones, were carried out. There were some minor changes in plasma constituents in the high-dose group; however, thyroid hormones were not affected, except in the 2-thiouracil group. No significant histological changes were observed in any of the organs of the rats fed LS 1 (Parish, 1986d).

In a study comparing the toxicity of two types of smoke flavourings, 5 groups of 20 male and 20 female rats were maintained for 90 days on diets containing 0, 0.3, or 3.0% of a basic aqueous condensate or 0.025 or 0.25% of an oil soluble smoke flavouring derived from the tar-like precipitate of a freshly prepared condensate. Ten animals died during the course of the study, but death was not related to the administration of test compound. Weight gain was similar in test and control animals. No haematological changes were observed at week 5 or at termination of the study. At termination of the study, organ-weight analyses and gross pathological and histopathological studies of the major organs showed no compound-related effects (Eschenberger, 1963).

Long-term studies
No information available.

Observations in man
No information available.

LIQUID SMOKE FLAVOURINGS (TARRY EXTRACT)

BIOLOGICAL DATA

Biochemical aspects

No information available.

Toxicological studies

Many of the studies summarized here were carried out using a liquid smoke flavouring preparation derived from the tarry fraction of smoke condensate prepared from alder wood (Alnus incana). Diethyl ether was used for the extraction process. Seventy-seven substances have been identified in the smoke flavouring. The major components were phenols and their alkylated derivatives (alkylated phenols, guaicols, and siringols). PAHs were present below 10 ppb, and no nitrosamines were detected (limit of detection, 5 ppb) (Miler, 1978).

Special studies on mutagenicity

A smoke flavouring derived from hickory was negative in the Ames Salmonella microsome plate test, with or without activation, at levels of 0.005 µl to 10 µl per plate (Jagannath & Brusick, 1979).

The smoke flavouring preparation from alder wood was tested using Salmonella typhimurium strains TA98, TA100, TA1535, and TA1537, with or without metabolic activation, at dose levels ranging from 28.3 to 2292 µg per plate. No mutagenic effects were observed (Jensen, 1986).

Acute toxicity[1]

Species	Sex	Route	LD_{50} (g/kg b.w.)	Reference
Mouse (Swiss albino)	M	oral	2.8	Fitko, 1979a
	F		2.3	
Rat (Wistar)	M		4.0	Fitko, 1979b
	F		3.5	
Pig	M & F		3.6	Fitko, 1979c

[1] The tested substance was the smoke flavouring preparation from alder
wood.

Short-term studies

Mice

Swiss albino mice were maintained for 90 days on diets
containing 0, 0.1, 0.5, 1.0, or 2.0% of the smoke flavouring derived
from alder wood. The number of mice was 35 females and 25 males in
each of the control and high-dose groups and 12-14 of each sex in each
of the intermediate groups. Eleven animals died during the course of
the study. Deaths occurred in all groups and were not compound-related.
Body-weight gains in all but the lowest-dose animals were increased
over controls, and were associated with increased food intake.
Haematologic studies, urinalysis, and clinical chemistry studies at
weeks 6 and 13 showed no significant differences between the control
and high-dose animals. At autopsy, organ-weight analysis showed
significant increases in weights of the kidneys and thyroid.
Histological examination showed compound-related effects in the liver,
kidneys, gastrointestinal tract, lungs, and spleen in the test groups,
with the exception of the 0.1% group. The effects that were noted
included parenchymatous degeneration of the renal tubules of the
kidneys, oedema and hyperaemia of the spleen, parenchymatous

degeneration of the liver, and oedema and hyperaemia in the mucosa of the stomach and the submucosa of the small intestine (Fitko, 1979d).

NOTE: The pathology terms used in this study and all the other studies authored by Fitko that are summarized in this monograph are those of the author. No criteria were available for the terms used in the pathology reports, hence complete evaluations of these studies are difficult because of the lack of definitions of the terms used.

Rats

Four groups, each containing 15 male and 15 female Wistar rats, were fed diets containing 0, 0.1, 0.3, or 1.0% of the tarry fraction from a smoke flavouring preparation for 13 weeks. Weight gain was significantly decreased in the high-dose group. At autopsy, organ-weight analysis showed a significant increase in thyroid weight of both male and female rats in the high-dose group. The thyroid glands of animals in the high- and medium-dose groups were noticably enlarged, compared to the controls. Histological examination indicated dose-related changes in the thyroid, characterized by areas of atrophy and large acini, a condition similar to colloidal goitre in man (Hercules, 1977).

Five groups of equal numbers of male and female Wistar rats were maintained for 90 days on diets containing 0, 0.1, 0.5, 1.0, or 2.0% of the smoke flavouring derived from alder wood. The number of rats was 35 of each sex in each of the control and high-dose groups and 15 of each sex in each of the intermediate groups. Haematologic, serum, clinical chemistry, and urine analyses carried out at 0, 6, and 13 weeks showed no significant differences between the control and high-dose animals. Body-weight gains were similar in test and control animals, except for the high-dose group in which the females showed a decrease, and males an increase, in body-weight gain compared to controls. Food consumption was higher in the high-dose group than in the controls. At autopsy, organ-weight analysis showed significant increases in liver, kidney, and heart weights, except in animals in the low-dose group. Macroscopic examination of the organs showed the following compound-related changes: catharrhal inflammation of the mucosa and small intestine and hyperaemia of the mucosa of the stomach in dosed groups, with the exception of the low-dose group. Histopathological examination revealed, in addition to changes in the gastrointestinal tract, changes in the liver (parenchymal degeneration and hyperaemia), kidneys (parenchymal degeneration of the renal glomerulus), and lungs (hyperaemia). No significant effects were observed in the low-dose group (Fitko, 1979e).

Pigs

Five groups, each containing 3 male and 3 female "large white" pigs, were administered in the diet the smoke flavouring derived from alder wood at dose levels equivalent to 0, 200, 600, 1000, or 1400 mg/kg b.w./day for 90 days. Each dose was given to the animals in 2 portions. Body-weight gains showed significant decreases with increasing dietary concentration of the test compound. However, this was associated with decreased food intake. Haematologic, serum, clinical chemistry, and urine analyses of the control and high-dose animals at weeks 0, 6, and 13 showed no consistent compound-related effects, with the exception of a dose-related increase in serum bilirubin. A liver function test (BSP retention) of high-dose and control animals at week 13 showed no significant differences. At autopsy, organ weight analysis showed no compound-related effects except for the liver and the uterus. In the case of the liver, there was a significant decrease in weight in the lower-dose groups, and an increase in weight in the high dose groups. The uterus showed a dose-related increase in weight. Macroscopic examination of the organs showed inflammation of the stomach and intestines, as well as changes in the kidneys and lungs. Histological examination of the tissues showed dose-related changes in the gastrointestinal tract, ranging from inflammation to prominant proliferation of lymphoid tissue into the submucosa of the stomach and small intestines and oedema of the mucosa in these organs. In addition, parenchymal degeneration of the liver and renal glomeruli and hyperaemia in the liver and lungs were observed. At the lowest dose fed, the only significant effects were inflammation of the stomach mucosa and lymphoid cell infiltration in the submucosa (Fitko, 1979f).

In another study, four groups of 3 male and 3 female pigs were fed diets containing the smoke flavouring derived from alder wood for 52 weeks at doses equivalent to 0, 200, 600, or 1000 mg/kg b.w./day. The daily dose was given to the test animals in two portions. There were no compound-related deaths during the course of the study. There was a dose-related decrease in mean body weights, and this was associated with decreased feed efficiency. Haematologic studies, serum clinical chemistry studies, and urinalyses were performed at 0, 1, 3, 6, and 12 months on animals in the control and high-dose groups. Serum cholesterol and direct bilirubin showed dose-related increases in the test animals. There were sporadic variations in haematologic parameters and certain serum enzymes (aspartate aminotransferase and alkaline phosphatase). Liver function tests (BSP retention) at 0, 6, and 12 months were similar at all test periods. At termination of the study, organ-weight analysis showed significantly increased relative weights of the liver and brain and decreased relative weights of the thymus in the high-dose group, and increased relative kidney weights in the intermediate group. Macroscopic examination of the organs showed inflammation of the gastrointestinal tract of all treated animals, and liver necrosis in the high-dose group. Histopathology studies showed changes ranging from inflammation of the gastrointestinal tract in all animals to liver necrosis in the high-dose group. Effects on the kidneys, which included parenchymal degeneration, hyperaemia, and necrosis, appeared to be most severe in the high-dose group (Fitko, 1979g).

Long-term studies
Mice
Groups of equal numbers of male and female Swiss albino mice were maintained for 16 months on diets containing smoke flavouring derived from alder wood at dose levels equivalent to 0, 150, 920, or 1650 mg/kg b.w./day for females, and 0, 150, 760, or 1720 mg/kg b.w./day for males. The control group and highest-dose group consisted of 60 mice of each sex, and the intermediate doses consisted of 25 mice of each sex. One hundred thirty-three of the 340 animals died during the course of the study. There were no significant differences between

survival in control and test groups. Gross pathology of 68/133 mice autopsied showed inflammation of the gastrointestinal tract in 34 animals and liver degeneration in 33 animals. These and other pathological changes were not dose related. No tumours were found in animals dying before sacrifice. Weight gain was slightly reduced in the high-dose group compared to the control and appeared to be associated with a decrease in feed efficiency. Haematology, serum, clinical chemistry studies, and urinalysis were carried out at 1, 3, 6, 12, and 16 months. There were no significant compound-related changes. At termination of the study, organ-weight analysis showed an increase in liver weights and a decrease in thyroid weights. Changes in weights of other organs appeared to be sporadic and not dose related. Histological studies showed a dose-related increase in the frequency of fatty degeneration of the liver, with liver necrosis in the highest-dose group. Inflammation of the gastrointestinal tract was observed in all groups. Other effects observed in the high-dose group included lymphocytic infiltration in the adrenals and lungs and hypertrophy of mucosa in the uterus. The number of test animals with tumours did not differ significantly from the number in the control group (the total number of animals examined in each group ranged from 19 to 24) (Fitko, 1979h).

Rats

Groups of rats were maintained for 24 months on test diets containing smoke flavouring derived from alder wood at dose levels equivalent to 0, 48, 260, or 630 mg/kg b.w./day for males and 59, 360, or 850 mg/kg b.w./day for females. The high-dose group consisted of 118 male and 117 female rats, the intermediate groups of 45 male and 45 female rats, and the control group of 117 males and 118 females. During the study 201 animals died; mortality was equally distributed among all groups. Lung disease followed by inflammation of the gastrointestinal tract and degeneration of the liver were listed as causes of death. No tumours were reported. Weight gain was lower in all test groups compared to controls and was associated with decreased feed efficiency. Haematology, serum clinical chemistry analyses, and urinalyses were carried out at 1, 3, 12, 18, and 24 months. There were sporadic changes in various parameters, but these were not dose related, and were within normal ranges. At termination of the study organ-weight analysis showed weight increases of the uterus, thyroid, lungs, and kidneys in females in the high-dose group, and in the hypophysis and thymus of males in the high-dose group. Histological studies of the tissues showed inflammation of the gastrointestinal tract, and necrosis of the liver and kidney in the two high-dose groups. There were no significant differences in the frequency distribution of any type of observed neoplasms between test and control animals. The data were evaluated using the methods described by Peto (1974) (Fitko, 1979i).

Observations in man

No information available.

COMMENTS

Toxicity data are available for liquid smoke flavourings derived from the aqueous and tar fractions.

The fractions were not mutagenic in bacterial systems (Salmonella typhimurium and E. coli strains), with or without activation. However, one liquid smoke preparation (aqueous fraction) which was inactive in the Salmonella test caused a significant increase in mutation frequency in the TK6 lymphoblast system. This compound was not positive ·in a lung tumour biossay system utilizing newborn mice, although toxic effects were observed.

Short-term studies were available for several types of liquid smoke flavourings (aqueous extract). The major adverse effects noted were significant increases in liver and kidney weights of the test animals. In one study, these changes were shown to be reversible. Only minor histological changes were reported in these animals. Minor changes in serum chemistry have also been reported. In these studies the no-effect level ranged from 0.25% to 3% of the liquid smoke preparation in the diet. It was also noted that in some studies there were significant increases in thyroid weights in the high-dose groups. However, changes in serum levels of T_3 and T_4 were reported in only one study.

Almost all the studies with smoke flavourings (tarry extract) relate to a single product. This product was tested in a number of species (mice, rats, and pigs). The tests include long-term studies in mice and rats. The major effects observed in the high-dose groups, in all species tested, were decreased body-weight gain and increased liver and kidney weights. Histopathological changes were primarily related to the gastrointestinal tract, liver and kidneys that included inflammation of the gastrointestinal tract and necrosis of the liver and kidney. These effects were observed in the groups fed the highest levels of the test material, and were minimal or absent in the low-dose groups. In lifetime feeding studies in mice and rats, there were no compound-related increases in tumour incidence or type in the treated animals. However, the lack of any neoplastic foci of the liver in rats or mice is unusual in studies of this duration. A complete evaluation of these studies is difficult because of the lack of definition of the terms used in the pathology report. High levels of smoke flavourings (tarry extracts) in the diets of rats also resulted in a significant increase in thyroid weights, and in one case histological changes were observed.

EVALUATION

The Committee viewed the use of smoke flavourings generically, keeping in mind that smoke flavourings are a replacement for traditional smoking practices, and as such they represent a definite improvement in that a large number of potentially toxic compounds are eliminated in their production. A similar view has been expressed by the Council of Europe (Resolution AP185/2).

The Committee concluded that so complex a group of products might not be amenable to the allocation of an ADI, and that smoke flavourings of suitable specifications could be used provisionally to flavour foods traditionally treated by smoking. However, as the safety data for these products were limited, new or novel uses of smoke flavourings should be approached with caution.

The Committee concluded that detailed information on the production and composition of smoke flavourings is required, and that it would be desirable to have further safety studies on a well-defined spectrum of smoke flavourings.

REFERENCES

Braun, A.G., Busby, W.F. Jr., Jackman, J., Halpin, P.A., & Thilly, W.G. (1986). Commercial hickory smoke flavouring is a human lymphoblast but does not induce lung adenoma in newborn mice. Food Chem. Tox. (in press).

Eschenberger, A.B. (1963). Biological examination of food additives. Unpublished report No. A-675 from Engineering Experiment Station, Georgia Institute of Technology, Atlanta, GA, USA. Prepared for Dunn Laboratories, Atlanta, GA, USA.

FCT (1965). Summaries of toxicological data, artificial smoke flavourings, II. "Liquid smoke flavour". Safety evaluation by oral administration to rats for 90 days. Fd. Cosmet. Tox., 3, 145-147.

Fitko, R. (1979a). Smoke Extract Flavour. Study of acute oral toxicity in the mouse. Unpublished report No. 2 from the Institute of Veterinary Sciences, Olsztyn, Poland. Submitted to WHO by Broste Industri, Copenhagen, Denmark.

Fitko, R. (1979b). Smoke Extract Flavour. Study of acute oral toxicity in the rat. Unpublished report No. 3 from the Institute of Veterinary Sciences, Olsztyn, Poland. Submitted to WHO by Broste Industri, Copenhagen, Denmark.

Fitko, R. (1979c). Smoke Extract Flavour. Study of acute oral toxicity in the pig. Unpublished report No. 4 from the Institute of Veterinary Sciences, Olsztyn, Poland. Submitted to WHO by Broste Industri, Copenhagen, Denmark.

Fitko, R. (1979d). Smoke Extract Flavour. Ninety-day oral toxicity study in the mouse. Unpublished report No. 5 from the Institute of Veterinary Sciences, Olsztyn, Poland. Submitted to WHO by Broste Industri, Copenhagen, Denmark.

Fitko, R. (1979e). Smoke Extract Flavour. Ninety-day oral toxicity study in the rat. Unpublished report No. 6 from the Institute of Veterinary Sciences, Olsztyn, Poland. Submitted to WHO by Broste Industri, Copenhagen, Denmark.

Fitko, R. (1979f). Smoke Extract Flavour. Ninety-day oral toxicity study in the pig. Unpublished report No. 7 from the Institute of Veterinary Sciences, Olsztyn, Poland. Submitted to WHO by Broste Industri, Copenhagen, Denmark.

Fitko, R. (1979g). Smoke Extract Flavour. Twelve-month oral toxicity study in the pig. Unpublished report No. 10 from the Institute of Veterinary Sciences, Olsztyn, Poland. Submitted to WHO by Broste Industri, Copenhagen, Denmark.

Fitko, R. (1979h). Smoke Extract Flavour. Sixteen-month oral toxicity study in the mouse. Unpublished report No. 8 from the Institute of Veterinary Sciences, Olsztyn, Poland. Submitted to WHO by Broste Industri, Copenhagen, Denmark.

Fitko, R. (1979i). Smoke Extract Flavour. Twenty-four oral toxicity study in the rat. Unpublished report No. 9 from the Institute of Veterinary Sciences, Olsztyn, Poland. Submitted to WHO by Broste Industri, Copenhagen, Denmark.

Hercules (1977). Herocsef smoke extract flavour. Three-month oral toxicity study in Wistar rats. Unpublished report No. CL76,107,1325, from Hercules Powder Co. Ltd., London, England. Submitted to WHO by Unilever Research and Engineering, Shambrook, England.

Jagannath, D.R. & Brusick, D.J. (1979). Mutagenicity evaluation of natural hickory smoke flavour - Code 402. Unpublished report No. 20988 from Litton Bionetics, Inc., Kensington, MD, USA, for Stange Company, Chicago, IL, USA.

Jensen, J.C. (1986). Ames Salmonella/microsome assay with Scansmoke concentrate (SEF). Unpublished report from Scantox, Denmark. Submitted to WHO by Broste Industri, Copenhagen, Denmark.

Lijinsky, W. & Shubik, P. (1965). Summaries of toxicological data, artificial smoke flavour, the detection and estimation of polycyclic hydrocarbons in liquid smoke. Fd. Cosmet. Tox., 3, 145-147.

Miler, K. (1978). Smoke extract flavours, description of the product to be tested in toxicity studies on the mouse, the rat and dog. Unpublished report prepared by the Polish Meat and Fat Research Institute, Warsaw, Poland. Submitted to WHO by Broste Industri, Copenhagen, Denmark.

Mortelmans, K.E. & Eckford, S. (1980). Microbial mutagenesis testing of substances, compound report F76-054, F76-055 and F6-076. Unpublished report No. LSV-6909 from S.R.I. International.

Parish, W.E. (1986a). Ingestion toxicity of liquid smokes, Part 2. Four-week rat feeding trial with Oss liquid smoke. Unpublished report No. PES 82,1022 from Unilever Research and Engineering. Submitted to WHO by Unilever Research and Engineering, Shambrook, England.

Parish, W.E. (1986b). Ingestion toxicity of liquid smokes, Part 3. Four-week rat paired feeding trial with Oss and Zwan liquid smoke. Unpublished report No. PES 85,1005 from Unilever Research and Engineering. Submitted to WHO by Unilever Research and Engineering, Shambrook, England.

Parish, W.E. (1986c). Ingestion toxicity of liquid smokes. Thirteen-week rat feeding trial with Oss Liquid smoke. Unpublished report No. D86/057 from Unilever Research and Engineering. Submitted to WHO by Unilever Research and Engineering, Shambrook, England.

Parish, W.E. (1986d). Ingestion toxicity of liquid smokes, Part 1. Four-week rat feeding trial with Zwan liquid smoke. Unpublished report No. PCW 79,1294 from Unilever Research and Engineering. Submitted to WHO by Unilever Research and Engineering, Shambrook, England.

Peto, R. (1974). Guidelines on the analysis of tumour rates and death rates in experimental animals (editorial). Brit. J. Cancer, 29, 101-103.

Raltech (1981). Ames Salmonella/microsome mutagenicity tests, a smoke product. Unpublished report No. 862001 from Raltech Scientific Services, Madison, WI, USA for Red Arrow Products Co., Manitowoc, WI, USA.

WARF (1961). Subacute toxicity study on a liquide smoke flavour. Unpublished report from Wisconsin Alumini Research Foundation. Submitted to WHO by Red Arrow Products, Co., Manitowoc, WI, USA.

WRC (1963). Liquid smoke flavour. Safety evaluation by oral administration to rats for 90 days. Unpublished report from Woodard Research Corporation for the Griffith Lab. Inc.

FOOD COLOURS

BEET RED AND BETANINE

Beet red is the colour obtained from the red beetroot, the principal component of which is betanine. This food colour was last reviewed at the twenty-sixth meeting of the Committee (Annex 1, reference 59), when the previously allocated temporary ADI "not specified" was withdrawn because the additional information required by the Committee at its eighteenth and twenty-second meetings was not available, that is, data on metabolism and long-term toxicity (Annex 1, references 35 and 47).

Since the previous evaluation, additional data have become available and are summarized and discussed in the following monograph.

BIOLOGICAL DATA

Biochemical aspects

Absorption, distribution, metabolism, and excretion

When betanine (4.5 μmole) was injected i.v. to rats, urinary excretion was rapid, 88% of the dose appearing in urine within 4 hours, and the plasma half-life was 32 minutes. Orally-administered betanine was poorly absorbed in rats and most of the dose was metabolized in the gastrointestinal tract; approximately 3% of the oral dose appeared in urine and a similar amount in faeces (Krantz et al., 1980).

Toxicological studies

Special studies on carcinogenicity

Rats

In a two-generation study, rats were given 50–78 mg betanine/kg b.w./day in drinking water throughout their lives. No evidence of carcinogenicity was reported (Druckrey, 1959).

No increase in tumours was observed in rats given repeated doses of betanine by subcutaneous injection (Druckrey, 1959).

A short-term study was performed to assess the ability of beet red to initiate or promote hepatocarcinogenesis in rats. Groups of female Sprague-Dawley rats (6–11 animals/group) were partially hepatectomized and treated with four different beet pigment preparations to assess their ability to initiate carcinogenesis: fermented betacyanin solution (50 mg/kg), pure betanine (50 mg/kg), degraded betanine (50 mg/ kg), or a diet containing 2000 mg betacyanin/kg. N-Nitrosodiethylamine (10 mg/kg) was used in a positive control group. Another group previously initiated with N-nitrosodiethylamine was given a betacyanin solution (100 ppm, equivalent to 3.5 mg/rat/day) to determine the ability of betacyanin to promote carcinogenesis after initiation relative to control and phenobarbitone-pretreated rats. After 6 months (promotion studies) or 8 months (initiation studies) the livers were examined histologically and histochemically for gamma-glutamyl peptidase foci. There was no evidence that betalain preparations initiated or promoted hepatocarcinogenesis (Schwartz et al., 1983).

Special studies on mutagenicity

Beet red was found to be non-mutagenic against 5 strains of Salmonella typhimurium in the Ames test, with or without metabolic activation by S-9 preparations, at concentrations of 500–2500 µg/plate (von Elbe & Schwartz, 1981). At higher concentrations (50 mg/plate), beet red was reported to be weakly mutagenic against S. typhimurium, with or without metabolic activation (Ishidate et al., 1984).

No mutagenic activity was detected in studies using
Escherichia coli or S. typhimurium assays, with or without metabolic
activation by rat S-9 preparations, or intestinal microbial
preparations. No DNA damage was detected in the E. coli rec assay
(Haveland-Smith, 1981).

Beet red did not induce chromosomal aberrations in Chinese
hamster fibroblast cells in culture (Ishidate et al., 1984).

Acute toxicity
Rats
No deaths were reported in rats given high oral doses of
beetroot red (Druckrey, 1959).

Single doses of betanine injected i.v. into anaesthetised
rats caused a transient increase in blood pressure and heart rate, the
effect of 0.9 μmole betanine being about equivalent to that of 2 nmole
adrenalin (Kranz et al., 1980).

Short-term studies
Rats
Groups of six rats were fed beet red preparations containing
2000 ppm betalains in the diet for 7 days. No significant differences
were noted in body-weight gain, food intake, or gross pathological
features relative to controls (von Elbe & Schwartz, 1981).

Long-term studies
(see "Special studies on carcinogenicity").

Observations in man
No information available.

COMMENTS AND EVALUATION
Previous Committees had considered beet red together with
its major colour component, betanine. This Committee decided that it
would be appropriate to evaluate these food colours separately and

pointed out that, for the compound betanine, insufficient data were available to establish an ADI, since the information available to the Committee did not meet currently accepted standards.

In evaluating beet red, the Committee took into account the principles laid down by the Committee at its twenty-first meeting (Annex I, reference 44) and endorsed in Annex III of "Principles for the Safety Assessment of Food Additives and Contaminants in Food" (Annex 1, reference 76). Thus, when the concentrate is used to enhance the colour of beet products, it could be considered as food. If, on the other hand, the concentrate is used more generally as a colourant, careful specifications need to be established. Because nitrate is a component of beet red, it is necessary to ensure that levels of nitrate do not exceed the specifications. Under these conditions beet red could be used according to good manufacturing practice with an ADI "not specified", keeping in mind the need to limit the nitrate content of foods produced for infants and young children.

REFERENCES

Druckrey (1959). Unpublished data submitted to WHO.

von Elbe, J.H. & Schwartz, S.J. (1981). Absence of mutagenic activity and a short-term toxicity study of beet pigments as food colourants. Arch. Toxicol., 49, 93-98.

Haveland-Smith, R.B. (1981). Evaluation of the genotoxicity of some natural food colours using bacterial assays. Mutation Res., 91, 285.

Ishidate, M., Sofuni, T., Yoshikawa, K., Hayashi, M., Nohmi, T., Sawada, M., & Matsuoka (1984). Primary mutagenicity screening of food additives currently used in Japan. Fd. Chem. Toxicol., 22, 623.

Krantz, C., Monier, M., & Wahlstrom, B. (1980). Absorption, excretion, metabolism and cardiovascular effects of beetroot extract in the rat. Fd. Cosmet. Toxicol., 18, 363-366.

Schwartz, S.J., von Elbe, J.H., Pariza, M.W., Goldsworthy, T., & Pitot, H.C. (1983). Inability of red beet betalain pigments to initiate or promote hepatocarcinogenesis. Fd. Chem. Toxicol., 21, 531-535.

CANTHAXANTHIN

EXPLANATION

The Committee was aware that canthaxanthin has been used as a direct food additive, as a feed additive, and as an orally-administered pigmenting agent for human skin in both pharmaceutical and cosmetic applications. It was evaluated for acceptable daily intake at the tenth and eighteenth meetings of the Committee (Annex 1, references 13 and 35).

The present Committee was asked to review the safety of canthaxanthin as a food additive because of reports of crystalline deposition in the retina during its use as an orally-administered skin pigmenting agent. The dose that resulted in this deposition was within the ADI established by the Committee at its eighteenth meeting (Annex 1, reference 35).

Canthaxanthin is a diketo carotenoid pigment with an orange-red colour. It occurs in the edible mushroom, chanterelle (Cantharellus cinnabarinus), in the plumage and organs of flamingoes, the scarlet ibis (Guara rubra), and the roseate spoonbill (Ajaja ajaja), and in various crustacea and fish (trout, salmon) (Haxo, 1950; Fox, 1962a, b; Thommen & Wackernagel, 1963).

BIOLOGICAL DATA

Biochemical aspects

Absorption, distribution, and excretion

Rats

Adult rats were fed a range of oral doses of canthaxanthin (not specified) for 13 and 20 weeks, respectively. Canthaxanthin accumulated in fat and some organs, notably the liver and spleen. Only slight depletion of canthaxanthin from fat occurred over a period of 1 month, indicating very slow elimination from this tissue (Hoffmann-La Roche, 1986).

Groups of rats were given daily doses of 0.6, 6, or 60 mg canthaxanthin/kg b.w. daily for five weeks. Highest organ concentrations were found in the liver and spleen, the tissue levels corresponding to the three dietary dose levels being 0.9, 12, and 125 µg/g liver, and 2.6, 50, and 67 µg/g spleen, respectively. Levels in other organs were much lower (0.2-1.5 µg/g at the highest dose level). After discontinuing administration of canthaxanthin, tissue levels in the adrenals and small intestine fell to one-quarter to one-third of their original levels over a period of 2 weeks (adrenals) or 1 month (intestine) (Hoffmann-La Roche, 1986).

When rats were fed daily doses of 50-60 mg canthaxanthin/kg b.w. for 9 weeks, the concentration in the eye remained at approximately 0.1 µg/g with no further accumulation. After administration of doses of 1.2, 2.0, 3.4, 5.6, 9.8, 16.7, or 28.4 mg canthaxanthin/kg feed (equivalent to 1.4 mg/kg b.w./day at the top-dose level) for 20 weeks, maximum concentrations in the eye were found to be about 0.01 µg/g; the residual levels in the eyes fell to 0.002 µg/g over a 4-week period after removal of canthaxanthin from the diet (Hoffmann-La Roche, 1986).

Chickens

Canthaxanthin earlier was reported not to exhibit provitamin A activity (Hoffmann-La Roche, 1966). However, in recent studies in chickens, canthaxanthin was shown to be converted to vitamin A in small amounts. Groups of 15 broiler chickens, 37 days old, were given diets containing 8.9, 18, or 35 mg canthaxanthin/kg feed along with 0, 300, or 600 i.u. vitamin A/kg feed. At each dietary level of vitamin A, canthaxanthin caused a dose-related increase in plasma and liver concentrations of both vitamin A and canthaxanthin. The lowest level of canthaxanthin (8.9 mg/kg feed), in the absence of dietary vitamin A, yielded higher plasma and liver vitamin A levels than did a diet containing 600 i.u. vitamin A/kg in the absence of canthaxanthin (Hoffmann-La Roche, 1986).

Distribution studies with radiolabelled canthaxanthin (dose not specified) fed to laying hens resulted in deposition of 30-40% of the dose in the egg yolk and 7% in body tissues; 42-54% was excreted. During a 10-day period of depletion after withdrawal of canthaxanthin, the body stores were mobilized and excreted via eggs and excrement. The canthaxanthin concentration in the fat was reported to be low and to remain constant (Hoffmann-La Roche, 1986).

Dogs

The distribution of canthaxanthin in the tissues was investigated in dogs which had received 50, 100, or 250 mg/kg b.w./day for 52 weeks, corresponding to total doses of 200, 400, or 1,100 g respectively. The highest mean tissue concentration was seen in adipose tissue (24 µg/g in the top-dose group). Relatively high concentrations were also seen in adrenals (15.1 µg/g), skin (9.62 µg/g) and liver (8.1 µg/g) in the low-dose group. The total amount of canthaxanthin extracted from 8 eyes of treated dogs was 0.1-0.4 µg, but ophthalmological examinations were not performed, so it was not possible to determine whether crystalline deposits had formed (Hoffmann-La Roche, 1986).

Humans

A pharmacokinetic study was performed in which plasma levels of canthaxanthin were measured at intervals after multiple dosing of human volunteers. Ten subjects were each given 1 mg canthaxanthin

6 times a day for 5 days, corresponding to a total dose of 30 mg. A further ten subjects received 8 mg canthaxanthin 6 times a day for 2 days, total dose 96 mg. Blood samples were taken at the start and at 12-hour intervals for 8 days, and canthaxanthin concentrations were determined by HPLC. The elimination half-life was calculated as 4.5 days in each group and the proportion of the dose absorbed was estimated to be 12 and 9%, respectively. The calculated steady state plasma concentrations of canthaxanthin after daily ingestion of 6 mg (6 times 1 mg) or 48 mg (6 times 8 mg) was calculated as 1843 or 10,346 µg/l, respectively (Kubler, 1986).

Toxicological studies
Special studies on carcinogenicity
Mice

When given orally, canthaxanthin exhibited no promotional activity in mice treated dermally with dimethylbenz(a)anthracene or benzo(a)pyrene (Mathews-Roth, 1982; Santamaria et al., 1982).

Oral doses of 6680 mg/kg b.w./day gave some protection against the skin carcinogenicity of regular exposure to UV radiation (Mathews-Roth, 1982).

Rats

Groups of rats were treated with canthaxanthin at doses of 0, 250, 500, or 1000 mg/kg/day for 104 weeks. A summary report stated that some changes in blood chemistry parameters were observed, mainly at the intermediate and high-dose levels, relating to a (non-specified) liver effect. No changes in tumour incidence were seen at up to 78 weeks of treatment, but after 104 weeks there was a slight, but non-dose related, increase in benign liver tumours in treated females. Full details of this study were not available; the report stated that the significance and nature of these findings will be evaluated in an additional study aimed at determining the no-effect level (Hoffmann-La Roche, 1986).

Special studies on ocular toxicity

Preliminary results of studies in rabbits on the ocular effects of canthaxanthin (dose not specified) did not reveal any deposits in the retina, but small alterations (prolongation) in dark adaptation were observed. The significance of these results is controversial (Hoffmann–La Roche, 1986).

Special studies on reproduction

A three–generation study using 0 or 1000 ppm canthaxanthin in the diet revealed no adverse effects in any generation (Hoffmann–La Roche, 1966).

No adverse effects on reproductive function were observed when doses of 0, 250, 500, or 1000 mg canthaxanthin/kg b.w./day were given throughout 3 generations (Hoffmann–La Roche, 1986).

Acute toxicity

Species	Route	LD_{50} (mg/kg b.w.)	Reference
Mouse	oral	10,000	Hoffmann–La Roche, 1966

Short-term studies
Dogs

Groups of three male and three female dogs received 0, 100, or 400 mg/kg b.w. canthaxanthin daily for 15 weeks. No significant effects were noted on body weight of control or test groups or on their general health (Hoffmann–La Roche, 1966).

Long-term studies
See also "Special studies on carcinogenicity"
Mice

In a preliminary outline report of an 80–week study in mice in which the animals received 0. 250. 500. or 1000 mg canthaxanthin/kg b.w./day, it was stated that no signs of systemic toxicity and no changes of any tumour incidence were seen that could be related to treatment (Hoffmann—La Roche, 1986).

Rats

Groups of 25-30 male and female rats received 0, 0.5, 2, or 5% canthaxanthin in their diets for 93-98 weeks. No adverse effects were noted on food consumption or weight gain. Mortality and tumour incidence were not increased (Hoffmann—La Roche, 1966).

Dogs

Oral doses of 0, 50, 100, or 250 mg canthaxanthin/kg b.w./day for 52 weeks were well tolerated and no adverse effects attributable to treatment were observed (Hoffmann—La Roche, 1986).

Observations in man

Six out of a group of 42 patients with a history of urticaria suffered a recurrence of their symptoms within 23 hours after an oral challenge with 410 mg canthaxanthin taken as three divided doses over 3 hours (Juhlin, 1981).

The ingestion of doses of about 30-120 mg canthaxanthin daily (approximately 0.4-1.7 mg/kg b.w./day) for 3 months to several years in medicinal or oral sun-tanning preparations has been associated with a retinopathy in some individuals characterised by glistening, golden crystals in the inner layers of the retina, up to 10-14 μm in size (Boudreault et al., 1983; Cortin et al., 1984; Ros et al., 1985). The crystalline deposits occur mainly in a ring between 5° and 10° around the fovea, less numerous in the fovea and rarely in the foveola (Cortin et al., 1982). Occasionally, deposits have been reported nasally of the disc (Metge et al., 1984) or scattered in the posterior fundus (cited in Daicker et al., 1987) and in one case only in the periphery of the fundus in the inner layer of a retinoschisis (Cortin et al., 1982). A total dose of 75-178 g canthaxanthin has been found to be effective in 50% of subjects and numerous cases have now been described (Franco et al., 1985; Hennekes et al., 1985; McGuiness &

Beaumont, 1985; Meyer et al., 1985; Philipp, 1985; Saraux & Laroche, 1983; Weber et al., 1985a,b; Weber & Goerz, 1985).

In most cases, pigment deposition is not associated with any detectable functional changes, but occasionally there have been complaints of dazzle or blurred vision (Cortin et al., 1984; Hennekes et al., 1985; Philipp, 1985); visual field defects have been described in only one report (Ros et al., 1985). The EOG is normal or subnormal and dark adaptation may be delayed; scotopic vision after exposure to glare is reduced while the ERG is normal or with b-wave changes (Boudreault et al., 1983; Metge et al., 1984; McGuiness & Beaumont, 1985; Weber et al., 1985b; Hennekes et al., 1985; Philipp, 1985).

Twenty-five patients were re-examined 2-10 months after therapy with canthaxanthin and β-carotene was discontinued. Dark adaptation and ERG had normalized, but the crystalline retinopathy and pigment epithelial defects showed no signs of reversibility (Weber & Goerz, 1986).

Increased susceptibility to retinal deposits has been associated with age and a number of clinical factors, including focal epitheliopathy, ocular hypertension, and possibly co-administration of β-carotene (Cortin et al., 1984).

In a biostatistical evaluation of 253 cases having received treatment with canthaxanthin, of whom 33 (15%) had retinal deposits, the median yearly dose in subjects free from visible retinal deposits was 5.3 g, whereas the corresponding figure for the group with pigment deposits was 14.4 g. The lowest dose at which deposits were recorded was 7 g canthaxanthin/year and no retinal deposits were found in patients receiving less than 30 mg/day (Hoffmann-La Roche, 1986).

The eyes of a female patient, aged 72 years, who had retinal deposits and who had died under anaesthesia, were examined by light and electron microscopy, and the extracted pigment was examined by mass and proton magnetic resonance spectroscopy. There were red, birefringent crystals in the inner layers of the entire retina, particularly large and numerous perifoveally where they were clinically visible. The crystals were located in the inner neuropil where an atrophy of the inner parts of the Muller cells was observed. The compound was identical to canthaxanthin and the retina contained up to 42 µg/g tissue besides a minor amount of other carotenoids. Of the other ocular tissue, only the ciliary body contained measureable amounts of canthaxanthin (Daicker et al., 1987).

Canthaxanthin was measured at autopsy in the tissues of 38 people, aged 22 to 96 years, none of whom were known to have received canthaxanthin therapeutically or in sun-tanning preparations. The tissues examined were mesenteric and sub-cutaneous fat, skin, liver, spleen, and blood serum. The highest concentrations were found in omentum and sub-cutaneous fat (mean concentrations, 0.2 and 0.3 µg/g, respectively). The mean concentrations in other tissues were: liver, 0.08 µg/g; skin and spleen, 0.04 µg/g; and serum, 0.024 µg/ml (Hoffmann-La Roche, 1986).

Fat samples from mesenterium and omentum and a liver sample were taken at autopsy from a 71-year-old woman who had died of bronchial carcinoma. The patient had previously ingested approximately 45 mg canthaxanthin/day for four years (total dose approximately 65 g) in a sun-tanning preparation. The concentrations of canthaxanthin in omentum and mesenteric fat were 270 µg/g and 158 µ/g respectively; lower levels of 5 µg/g were found in the liver (Hoffmann-La Roche, 1986).

Biopsy samples of orange-coloured fat (omentum) were obtained from two patients undergoing surgery. In one case the woman had taken a total dose of about 6 g canthaxanthin in a tanning preparation during 1983/84 and had stopped this intake 1-1.5 years before the biopsy; fat and serum canthaxanthin levels were 49 µg/g and 69 µg/l, respectively. In the second case the patient had taken a total dose of approximately 16 g over 2.5-3 years and the concentration in omentum was 34 µg/g (Hoffmann-La Roche, 1986).

COMMENTS

The Committee viewed the observation of crystal deposition in the retina as new data that warranted a complete review of this compound. Most of the new data from animal studies were available only as summaries, and therefore could not be used as a basis for evaluation. The Committee noted that, although pigment accumulation in the eye had been demonstrated analytically following oral administration of canthaxanthin to rats and dogs, no ophthalmoscopy had been carried out and no animal model for the human condition had been developed. In man, however, an estimate was made of the minimal level of exposure resulting in fundal pigment deposition in the retina.

In the light of all these considerations, the previous ADI was made temporary and reduced.

When setting an ADI, the Committee does not consider therapeutic use, which is a matter for clinical judgement. The cosmetic use of canthaxanthin as an orally-administered skin pigmenting agent was not anticipated when the ADI was established at the eighteenth meeting. Therefore, it is not included in the temporary ADI established at the present meeting; the temporary ADI applies only to the food and feed additive uses of canthaxanthin.

EVALUATION

Estimate of temporary acceptable daily intake for man

0-0.05 mg/kg b.w. (based on the minimal effect level for pigment deposition in the retina of human subjects, to which a 10-fold safety factor was applied).

Further work or information

Required

1. Details of the long-term studies in mice and rats for which summary data were submitted, including ophthalmological data where available.

2. Clarification of the factors that influence deposition in the eye, including the establishment of the threshold dose, the influence of dose and duration of exposure, the reversibility of pigment accumulation, and the investigation of potential animal models.

3. Clarification of whether pigment deposition is causally related to impaired ocular function.

REFERENCES

Boudreault, G., Cortin, P., Corriveau, L.A., Rousseau, A.P., Tardif, Y., & Malenfant, M. (1983). La rétinopathie à la canthaxanthine: 1. Etude clinique de 51 consommateurs. Can. J. Ophtalmol., 18, 325-328.

Cortin, P., Corriveau, L.A., Rousseau, A.P., Tardif, Y., Malenfant, M., & Boudreault, G. (1982). Maculopathie en paillettes d'or. Can. J. Ophtalmol., 17, 103-106.

Cortin, P., Boudreault, G., Rousseau, A.P., Tardif, Y., & Malenfant, M. (1984). La rétinopathie à la canthaxanthine: 2. Facteurs prédisposants. Can. J. Ophtalmol., 19, 215-219.

Daicker, B., Schiedt, K., Adnet, J.J., & Bermond, P. (1987). Canthaxanthin retinopathy: Light and electron microscopic investigation and physicochemical analysis. Graefe's Arch. Clin. Exp. Ophthalmol. (in press).

Fox, D.L. (1982a). Comp. Biochem. Physiol., 6, 1.

Fox, D.L. (1982b). Comp. Biochem. Physiol., 6, 305.

Franco, J.L., Adenis, J.P., Mathon, C., & Lebraud, P. (1985). Un nouveau cas de maculopathie en paillettes d'or. Bull. Soc. Ophtal. Fr., 85, 1035-1037.

Haxo, F. (1950). Botan. Gaz., 122, 228.

Hennekes, R., Weber, U., & Kuchle, H.J. (1985). Ueber canthaxanthin-schaden der netzhaut. Z. prakt. Augenheilk., 6, 7-9.

Hoffmann-La Roche (1966). Canthaxanthin. Unpublished report submitted to WHO by F. Hoffman-La Roche & Co., Basle, Switzerland.

Hoffmann-La Roche (1986). Canthaxanthin. Unpublished report submitted to WHO by F. Hoffman-La Roche & Co., Basle, Switzerland.

Juhlin, L. (1981). Recurrent urticaria: clinical investigation of 330 patients. Br. J. Dermatol., 104, 369-381.

Kubler, W. (1986). Biokinetic evaluation of canthaxanthin plasma levels after multiple doses of 1 mg and 8 mg canthaxanthin. Unpublished report submitted to WHO by F. Hoffman-La Roche & Co., Basle, Switzerland.

Mathews-Roth, M.M. (1982). Antitumour activity of beta-carotene, canthaxanthin and phytoene. Oncology, 39, 33-37.

McGuiness R. & Beaumont, P. (1985). Gold dust retinopathy after the ingestion of canthaxanthin to produce skin-bronzing. Med. J. Aust., 143, 622-623.

Metge, P., Maudirac-Bonnefoy, C., & Bellaube, P. (1984). Théraurismose rétinienne à la canthaxanthine. Bull. Mem. Soc. Fr. Ophtalmol., 95, 547-549.

Meyer, J.J., Bermond, P., Pournaras, C., & Zoganas, L. (1985). Canthaxanthin. Langzeiteinnahme und sehfunktion beim menschen. Dtsch. Apoth. Zeitung, 125, 1053-1057.

Philipp, W. (1985). Carotinoid-einlagerungen in der netzhaut. Klin. Mbl. Augenheilk., 187, 439-440.

Ros, A.M., Leyon, H., & Wennersten, G. (1985). Crystalline retinopathy in patients taking an oral drug containing canthaxanthin. Photodermatol., 2, 183-185.

Santamaria, L., et al. (1982). Médecine, biologie et environnement, 10, 359.

Saraux, H. & Laroche, L. (1983). Maculopathie à papillottes d'or après absorption de canthaxanthine. Bull. Soc. Opht. Fr., 83, 1273-1275.

Thommen, H. & Wackernagel, H. (1963). Isolation and identification of canthaxanthin in the lesser flamingo (Phoenicolnaias minor). Biochim. Biophys. Acta., 69, 387-396.

Weber, U. & Goerz, G. (1985). Augenscheden durch carotinoid-einnahme. Dtsch Aerzteblatt., 82, 181-182.

Weber, U., Goerz, G., & Hennekes, R. (1985a). Carotinoid-retinopathie: I. Morphologische und funktionelle befunde. Klin. Mbl. Auggenheilk., 186, 507-511.

Weber, U., Hennekes, R., & Goerz, G. (1985b). Carotinoid-retinopathie: II. Elektrophysiologische befunde bei 23 carotinoid-behandelten patienten. Klin. Mbl. Auggenheilk., 186, 507-511.

Weber, U. & Goerz, G. (1986). Carotinoid-retinopathie. III. Reversibilitat. Klin. Mbl. Auggenheilk, 188, 20-22.

CARBON BLACK

EXPLANATION

Activated carbon (synonyms, activated charcoal and decolorizing carbon), that is carbon black derived from vegetable material or lignites, was evaluated under the name "activated vegetable carbon (food grade)" at the fourteenth meeting of the Committee (Annex 1, reference 22). An ADI "not limited", except that good manufacturing practice be followed, was established. This refers to its use as a clarifying agent, not as a food colour. A toxicological monograph was published (Annex 1, reference 23), and new data evaluated at the present meeting are included in this monograph addendum.

BIOLOGICAL DATA

Four recent reviews have been published on carbon black (NCI, 1985; IARC, 1984; Rivin & Smith, 1982; NIOSH, 1978).

Biochemical aspects

Absorption, distribution, and excretion

Inhaled carbon black is retained in the lungs. Clearance is by macrophage uptake, retrograde mucociliary movement, and possibly secondary gastrointestinal ingestion. Absorption into the blood stream for systemic distribution evidently does not occur. There have been no reports of gastrointestinal absorption and carbon black is probably cleared in the faeces (Nau et al., 1962, 1976; NCI, 1985).

Toxicological studies

Special studies on the bioavailability of polynuclear hydrocarbons adsorbed on carbon blacks

In vitro studies

Samples of 3 different carbon blacks (rubber-grade oil furnace blacks, ASTM designation N-234, N-351, and N-375) were extracted with the following tissue fluids and cellular components: human plasma, swine serum, the supernatant of swine lung homogenates, and swine lung washings. All tissue fluids were poor eluters of benzo(a)pyrene (less than 0.005% of the absorbed benzo(a)pyrene content of the carbon black as determined by toluene extraction was extracted by the tissue fluids). Swine serum was a less efficient extractant than human plasma. Swine lung homogenate and lung washings were equally effective (Buddingh et al., 1981).

The extent of the elution of benzo(a)pyrene depends on the benzo(a)pyrene content of the carbon black and surface area of the particles, e.g. soot particles of 100 nanometers or less adsorb free benzo(a)pyrene from a protein medium (Falk & Steiner, 1952).

In vivo studies

Groups of 5 male and 5 female outbred CIGR mice were fed diets containing 0, 0.0082, 0.20, or 2.0% of 3 different carbon blacks (N-234, N-351 and N-371) for 3 successive generations. Litters from the F_0, F_1, and F_2 generations were killed at day 28, breeders from the F_2 generation were killed, and arylhydrocarbon hydroxylase activities of the liver and lungs were determined. Dietary exposure to carbon black had no effect on enzyme activity, suggesting that the elution of benzo(a)pyrene was not sufficient to induce increased enzyme activity in this system (Buddingh et al., 1981).

Special studies on carcinogenicity

Mice

Groups of 10-50 CFW white and/or CH_3 brown mice in a series of feeding experiments were administered, for periods of 12 to 18 months, either: (1) 10% carbon black; (2) 10% benzene-extracted carbon black; (3) benzene extract from carbon black; (4) 3-methyl-cholanthrene (MCA) or MCA adsorbed to flour; or (5) MCA adsorbed to benzene-extracted carbon black. The test material was dispersed in the basal diet by use of either a water-base mixture containing carboxy-methyl cellulose or an oil-base mixture containing cotton seed oil. For the control groups, the basal diet was supplemented with either the water-base mixture or oil-base mixture. At termination of the study, all the mice were killed and complete gross and microscopic examinations were made of all organs and tissues. No significant effects were observed in either the control groups or the groups given the unextracted carbon black. Mice fed extracted carbon black in the water-base diet developed a number of tumours (10/100). Nine of these tumours (3 intracutaneous fibrosarcomas, 3 begnin squamous papillomas, and 3 squamous metaplasas with malignancy) were considered to be due to benzene-extractive material that had not been completely removed from the benzene-extracted carbon black. No significant effects were reported in the group fed extracted carbon black in the oil-base mixture. Mice fed the benzene extract of carbon black in diets containing either the water-base or oil-base mixture developed tumours of the gastrointestinal tract and carcinomas of the stomach. In the groups of mice fed MCA, there was a high incidence of adenocarcinomas or squamous-cell carcinomas of the gastrointestinal tract. However, in the groups of mice fed MCA that was adsorbed to extracted carbon black, only 1 of 190 developed fibrosarcoma of the gastrointestinal tract (Nau et al., 1958).

Mice and rats

Groups of 24 or 48 Swiss mice or Harlan stock rats were administered, for more than 15 months, para-dimethylaminoazobenzene (DMAB), methyl cholanthrene, or 3,4-benzo(a)pyrene, either free or adsorbed onto various carbon blacks; carbon black alone was administered to other groups of mice and rats. The level of carbon black in the diet ranged from 9 to 18%. All animals were killed and necropsied, and selected tissues were examined histologically. No tumours were observed in the groups of mice or rats receiving carbon black only. In the group receiving free DMAB, 14/24 mice (58%) developed hepatic tumours. Of the groups receiving DMAB adsorbed onto carbon black, only one group developed tumours. The time to first tumour in this group was 10.25 months compared to 6 months in positive controls. None of the other groups treated with the adsorbed carcinogens developed tumours, although high incidences were observed in the test animals fed the carcinogens alone. Test animals treated with acetone suspensions of carbon black plus 3,4-benzo(a)pyrene developed a high incidence of tumours (54-69%); however, the onset of tumours was delayed when compared to positive controls (von Haam et al., 1958).

Groups of female 26-31 CF_1 mice and female 29-45 Sprague-Dawley rats were fed either 0 or 2.0 g carbon black (ASTM N-375) per kg of ground lab chow diet for 2 years. This dietary level was calculated to amount to an average consumption of 100 g/kg b.w./year for the mice and 38 g/kg b.w./year for the rats. (The average fat content of this rodent chow was later reported by these authors (1986) to be approximately 5% by weight.) Simultaneously, groups of mice and rats were exposed to carbon black for 52 weeks with or without the administration of 1,2-dimethylhydrazine (DMH) via 16 weekly i.p. injections at dose levels of 10 mg/kg b.w. in rats and 20 mg/kg b.w. in mice. Control animals were given the solvent (1mM EDTA) by injection. After 52 weeks or 2 years, the animals were killed, gross necropsies performed, and all lesions examined microscopically. The survival of all groups of animals was comparable; there was no apparent effect of carbon black ingestion on tumour incidence. A small non-significant incidence in colon tumours was seen in the group not treated with DMH in the 2-year study. In the 52-week groups in which carbon black plus DMH was administered, there were no enhancements in gastrointestinal, respiratory, mammary, or urinary tumours. However, groups given carbon black and DMH had an increased mortality (Pence & Buddingh, 1985).

Rats

Groups of 25 female Sprague-Dawley rats were administered carbon black (ASTM N-375) at a level of 0 or 2.0 g/kg diet. The study utilized a high-fat diet and consisted of 20% (w/w) corn oil added to a ground commercial chow diet. The average carbon black consumption was calculated to be 38 g/kg b.w./year. Colonic tumours were induced in the test groups by 16 weekly i.p. injections of DMH at 10 mg/kg b.w. All groups were maintained on test diets for 52 weeks, killed, subjected to necropsy, and all lesions were examined microscopically. Weight gain and food intake were not affected by any of the four regimens. DMH-treated rats had decreased survival due to intestinal tumours, and this effect was most prominent in the group also receiving carbon black. There were no colonic tumours in animals not treated with DMH. In the DMH group maintained on a high-fat diet, 60% of the females had colonic tumours. This was significantly different (P < 0.05) from the 76% seen in the DMH group maintained on the same high-fat diet which contained 2.0 g/kg carbon black (Pence & Buddingh, 1986).

Special studies on mutagenicity

Commercially produced furnace carbon black (rubber grade, CAS No. 1333-86-4) containing 194 ppm polynuclear aromatic hydrocarbons (PAHs) (determined on a benzene extract) showed limited toxicity but no mutagenetic activity in the following assays: (1) Salmonella assay, 5 tester strains of Salmonella typhimurium (TA98, TA100, TA1535, TA1537 and TA1538), with or without metabolic activation, at levels up to 7,500 µg/plate; (2) sister-chromatid exchange in Chinese hamster ovary cells, with or without metabolic activation, at test levels up to 1000 µg/ml; (3) mouse lymphoma cell L5178 assay, with or without metabolic activation, at test levels up to 15,000 µg/ml; (4) $C_3H/10$ T1/2 cell transformation assay at test levels up to 16,384 µg/ml; and (5) the Drosophila assay (Kirwin et al., 1981).

Short-term studies

No information available.

Long-term studies

Mice

Groups of 8-week old C_3H mice were fed diets containing either 0 or 10% thermal black for as long as 72 weeks. No significant gross or microscopic changes from normal were seen. This was a summary report and did not give any details (Nau et al., 1976).

Observations in man

No reports on oral ingestion of carbon black by humans were available. The available information relates to occupational exposure through inhalation. Three reviews have been published which examined the toxicity of carbon black to humans under these conditions (NIOSH, 1978; Rivin & Smith, 1982; IARC, 1984).

The major effect of carbon black in humans is on lung function. Other effects in humans attributed to carbon black are dermatological lesions, skin irritation, acute gastrointestinal diseases, myocardial dystrophy, and cardiovascular changes. In both of the reports on heart effects, there was concomitant exposure to carbon monoxide (Komarova, 1965, 1973, as cited by NIOSH, 1978).

IARC (1984) reviewed the available epidemiological data and concluded that the data provide inadequate evidence to evaluate the carcinogenicity of carbon black to humans.

COMMENTS

Carbon black used for colouring purposes falls within two main groups, those derived from hydrocarbons and those derived primarily from peat and plant materials, commercially described as vegetable black.

The food colouring uses of carbon blacks derived from both sources were evaluated by the Committee at the twenty-first meeting (Annex 1, reference 44). No ADI was established for food colouring uses from either source. A major concern of that Committee related to the question as to how strongly and irreversibly PAHs are adsorbed onto carbon black.

The present Committee considered data from studies involving carbon black prepared from hydrocarbon sources. Benzene extracts of certain carbon blacks were found to be carcinogenic to mice. These carcinogenic extracts contain PAHs adsorbed to carbon black. Data were available to show that only small amounts of PAHs (less than 0.005% of the benzene-extractable PAHs) were eluted from carbon black by biological fluids. Carbon black was not mutagenic in bacterial or mammalian systems. Dietary carbon black was not carcinogenic in limited lifetime studies in rats and mice at levels up to 10% of the diet. Information was also presented to show that carbon black was able to adsorb some chemical carcinogens and, under certain experimental conditions, was shown to reduce their carcinogenic potential.

No toxicological data were available on carbon black derived from vegetable sources.

EVALUATION
Carbon black (hydrocarbon sources)
Food contact materials
The use of carbon black from hydrocarbon sources is provisionally accepted in food contact materials, including wax coatings for cheese. Future specifications should include figures relating to residual PAHs.

Direct use in food
No ADI could be established (a) because carbon blacks from hydrocarbon sources have been shown to contain different amounts of known carcinogens and knowledge is lacking on the ability of man to extract such carcinogens upon ingestion and (b) because of limited feeding studies in experimental animals with defined carbon blacks.

Carbon black (vegetable black)
No ADI could be established because no toxicological data were available.

REFERENCES

Buddingh, F., Bailey, M.J., Wells, B., & Haesemeyer, J. (1981). Physiological significance of benzo(a)pyrene adsorbed to carbon blacks: Elution studies; AHH determinations. Amer. Ind. Hyg. Assoc. J., 42, 503-509.

Falk, H.L. & Steiner, P.E. (1952). The adsorption of 3,4-benzpyrene and pyrene by carbon blacks. Cancer Res., 12, 40-43.

von Haam, E., Titus, H.L., Caplan, I., & Shinowara, G.Y. (1958). Effect of carbon blacks on carcinogenic compounds. Proc. Soc. Exptl. Biol. Med., 98, 95-98.

IARC (1984). Monographs on the evaluation of the carcinogenic risk of chemicals to humans. International Agency for Research on Cancer. WHO/IARC Monograph, vol. 33, pp. 35-85.

Kirwin, C.J., Le Blanc, J.V., & Thomas, W.C. (1981). Evaluation of the genetic activity of industrially-produced carbon black. J. Tox. Environ. Health, 7, 973-989.

Nau, C.A., Neal, J., & Stembridge, V.A. (1958). A study of the physiological effects of carbon black - I. Ingestion. A.M.A. Arch. Ind. Health, 17, 21-28.

Nau, C.A., Neal, J., Stembridge, V.A., & Cooley, R.N. (1962). Physiological effects of carbon black - IV. Inhalation. Arch. Environ. Health, 4, 415-431.

Nau, C.A., Taylor, G.T., & Lawrence, C.H. (1976). Properties and physiological effects of thermal carbon black. J. Occup. Med., 18, 732-734.

NCI (1985). Monograph on human exposure to chemicals in the workplace: Carbon black. National Cancer Institute. NTIS publication No. PB86-152048, Springfield, VA, USA.

NIOSH (1978). Criteria for a recommended standard. Occupational exposure to carbon black. National Institute for Occupational Safety and Health. DHEW publication No. 78-204. US Department of Health, Education and Welfare, Washington, DC 20204, USA.

Pence, B.C. & Buddingh, F. (1985). The effect of carbon black ingestion on 1,2-dimethylhydrazine induced colon carcinogenesis in rats and mice. Tox. Letters, 25, 273-277.

Pence, B.C. & Buddingh, F. (1986). Co-carcinogenesis effect of carbon black ingestion with dietary fat on the development of colon tumours in rats. Manuscript submitted to Tox. Letters, September, 1986.

Rivin, D. & Smith, R.G. (1982). Environmental health aspects of carbon black. Rubber Chem. Technol., 55, 707-761.

CITRANAXANTHIN

EXPLANATION

Citranaxanthin is a synthetic compound. Its major present-day use is as an animal feed additive to impart a yellow colour to chicken fat and egg yolks. It can be used as colouring matter in the same manner and for the same purposes as other carotenoids (beta-carotene, beta-apo-8'-carotenal, beta-apo-8'-carotenoic acid ethyl ester, and canthaxanthin).

Citranaxanthin has about two-thirds of the vitamin A activity of beta-carotene in chickens. This is equivalent to 1,100 I.U. of vitamin A per 1 mg of citranaxanthin.

BIOLOGICAL DATA

Biochemical aspects

Absorption, distribution, and excretion

Twenty female Sprague-Dawley rats weighing about 170 g each received single i.v. doses of ^{14}C-citranaxanthin dissolved in polyethylene glycol. Each dose was equivalent to 1 mg/kg b.w. At 1, 24, 72, and 168 hours after administration five of the rats were sacrificed. The ^{14}C content of the following organs and tissues were determined: heart, liver, lungs, spleen, eyes, pituitary, brain, kidneys, pancreas, stomach, duodenum, thymus, periovarian and perirenal fats, ovaries, skin, hair, adrenals, thyroid, and muscle. The absolute amounts of radioactivity found in each organ were particularly high in the liver, lungs, and spleen. The radioactivity found in these organs ranged from 49% (1 hour) to 15% (168 hours) of the dose in the liver, 14% (1 hour) to 2.7% (168 hours) of the dose in the lungs, and 2.2% (1 hour) to 0.85% (168 hours) of the dose in the spleen.

The doses recovered in other organs after 1 hour were 0.7% in the heart, 0.1% in the brain, and 0.6% in the kidneys. Muscle and adipose tissue contained 0.3 and 2.6%, respectively of the dose administered after 1 hour, thus emphasizing the hydrophobic character of citranaxanthin. All other organs examined contained levels of radioactivity less than 0.1% of the dose/g 1 hour after administration. These levels, which were very low and were associated with large errors of measurement, decreased even further over the 168 hours after administration. The changes in levels of radioactivity with time were similar in all organs and tissues, decreasing slowly overall. A half-life of about 140 hours for this decrease was estimated by a logarithmic regression method (Morgenthaler, 1979).

Four female Sprague-Dawley rats weighing about 160 g each received $^{14}C-$ and ^3H-citranaxanthin orally by gastric tube. Citranaxanthin was administered as an oily suspension in carboxymethyl-cellulose. In both cases 95-98% of the radioactivity was recovered in faeces (69-80% (first day); 15-28% (second day); 0.7% (third day)). After 8 days, 0.2-0.8% of the radioactivity was found in the liver, 0.2% in the urine, and 0.5% in the carcass (Morgenthaler, 1978).

The average absorption rate of citranaxanthin in poultry was 50%. The rest was excreted in the faeces. About two-thirds of the absorbed amount was metabolized to vitamin A; about 40% was present in blood and other organs. One-third of the absorbed citranaxanthin was deposited in the fat of various tissues, in skin, and in egg yolk (Crina, 1974a,b).

Vitamin A activity of citranaxanthin

One hundred and twelve chickens (one-day old) were fed a diet free of vitamin A for 3 weeks. Divided groups received doses of 1.2, 2.4, 4.8, 6.0, 12.0, or 24.0 mg citranaxanthin daily over a period of five days. Liver analyses showed storage of citranaxanthin and vitamin A derived from citranaxanthin (Tiews, 1968a).

Chicks and quails received 40 ppm citranaxanthin and beta-carotene over 2 weeks in their diets after a 3-week rearing period without vitamin A. The vitamin A activity of citranaxanthin in chickens was higher than that of beta-carotene; in quails it was lower (Tiews, 1968b).

After a 21-day vitamin A-depletion period, groups of 20 laying hens were fed diets containing 1.5, 3, 6, or 10 ppm citran-axanthin. Colouration of egg yolk by citranaxanthin was observed. This effect was found to be influenced by dose and particle size (Kolk, 1974).

After a vitamin A-depletion period of 3 weeks, groups of 14 or 20 chickens were fed diets containing 8.6, 14, 24, or 40 ppm citranaxanthin for 2 weeks. The control groups received feed which contained the same amounts of beta-carotene instead of citranaxanthin. Determination of the liver vitamin A content showed that the formation of vitamin A from citranaxanthin depended upon the crystal size. The relative vitamin A activity was 66 to 94% of the activity compared with beta-carotene (Kolk, 1974).

The vitamin A activity of citranaxanthin was tested in chicken liver storage tests (Table 1). From this test and others it can be concluded that citranaxanthin has about two-thirds of the vitamin A activity of beta-carotene. This is equivalent to 1,100 I.U. of vitamin A/mg citranaxanthin (Kolk, 1974).

Table 1.   Vitamin A activity of citranaxanthin

Dosage (ppm in feed)	Vitamin A (I.U./liver)	Vitamin A (I.U./mg ingested carotenoid)	Relative vitamin A activity (β-carotene = 100%)
β-Carotene "water soluble"			
8.65	900	195	100
14.4	1,803	222	100
24.0	2,596	191	100
Citranaxanthin			
8.65	544	118	60
14.4	1,080	124	56
24.0	1,991	144	76

Toxicological studies

Special study on inhalation

Rats

Twelve animals were exposed to citranaxanthin crystalline/ air mixture at a concentration of 0.29 mg/liter of air during 8 hours. No deaths occurred. Toxic symptoms were not evident; only slight mucous membrane irritation was observed (BASF, 1972a).

Special study on mucous membrane irritation

Rabbits

Fifty mg citranaxanthin crystalline was applied to the eyes of rabbits. No irritation was seen. After 1 hour the eyes had a red appearence; after 24 hours and 8 days these symptoms were not noticeable (BASF, 1972a).

Special study on reproduction

Rats

A 3-generation reproduction study with groups of 20 male and 20 female Sprague-Dawley rats was conducted in which the animals were administered 1000, 3300, or 10,000 ppm citranaxanthin (8.6% dry powder, equivalent to 86, 284, and 860 ppm active ingredient). Another group was administered 9000 ppm dry powder without active ingredient. A control group remained untreated. Treatment of both males and females started 7 weeks before breeding and continued during the mating, pregnancy, and rearing periods. Fertility and reproduction performance were not influenced in any generations (F_0, F_1, or F_2) or groups. Mating, pregnancy, litter size, birth weights, and rearing were within normal limits. The indices of fertility, pregnancy, viability, and lactation did not differ between control and treatment groups. Animals of the F_2-generation were investigated for malformations; none were observed. Behaviour, appearance, feed and water intake, and weight gain were not affected by treatment in any generations or groups. No macroscopic pathological effects were observed in parents or their offspring. The final macroscopic examination of the F_2-generation animals after 9 weeks of age revealed no changes due to treatment. Comparisons of organ weights and histological examinations showed no differences among groups (Leuschner et al., 1976a).

Special study on skin irritation

Crystalline citranaxanthin was applied to the skin of rabbits in a 50% aqueous suspension for 20 hours. A red-brownish colouring was observed, but inflammation was not detected (BASF, 1972a).

Acute toxicity

Species	Route	LD_{50}[1] (mg/kg b.w.)	Reference
Mouse (10)	i.p.	> 6,400	BASF 1972a
Rat (10)	oral	> 6,400	BASF 1972a,b
Dog (6)	oral	> 1,590	Leuschner, 1976c

[1]   Higher doses could not be administered in these studies for technical reasons.

Short-term studies

Rats

Citranaxanthin was administered in the diet to groups of 13 to 15 male and female Sprague-Dawley rats at dose levels of 10, 20, 50, or 100 mg/kg/day for one month. Growth rates and feed consumption were not significantly different between the treated and the control groups during the test period. No toxic symptoms or abnormalities in urinalysis, biochemical values, blood analysis, wet organ weights, or histopathological changes were found (Kawase et al., 1972).

Four groups of 20 or 30 male and female Sprague-Dawley rats were maintained over a period of 91 days on a diet containing 25,000, 50,000, or 100,000 ppm 10% citranaxanthin dry powder (equivalent to 2,500, 5,000, and 10,000 ppm active ingredient). Citranaxanthin was tolerated without externally recognizable toxic symptoms and without impairment of feed ingestion or growth over a period of 91 days. A significant increase in total serum lipids was observed in both males and females during the treatment period. GPT values were temporarily increased after 8 weeks, but not after 12 weeks.

The absolute and relative weights of the liver and kidneys showed a significant increase when compared with those of the control group. All the increases were reversible in the post-observation period, except that the increase in kidney weight did not revert completely.

No salient pathological changes were observed in any goups. A few cases of enlargement of the liver proved after histological investigation to be hyperaemia, in some cases diffuse and in other cases patchy. Accumulated changes in the cylinders containing protein in the distal tubule sections of the kidneys were found in all groups (Hempel et al., 1973).

Long-term studies

Rats

Groups of 25 male and 25 female Sprague-Dawley rats were fed for two years with graded levels of 1000, 3300, or 10,000 ppm 8.6% citranaxanthin dry powder (equivalent to 86, 284, and 860 ppm active ingredient). Another group was administered 9000 ppm dry powder without active ingredient. A control group remained untreated. After 6 months the highest concentration of 10,000 ppm was raised to 20,000 ppm (1720 ppm active ingredient) and the concentration of the placebo was raised to 18,000 ppm for the remaining period of the experiment.

There were no signs of incompatibility in any groups throughout the experiment. Appearance, behaviour, weight gain, haematology, clinical chemistry, urinalysis, weight of organs, and their autopsy did not reveal any influence of treatment. There were no histological changes in organs or tissues. The number and kind of tumours or mortality did not differ among groups (Leuschner et al., 1976b).

Dogs

A 180-day feeding trial with 5 groups of 8 beagle dogs (4 males and 4 females) was performed with graded levels of 1000, 3300, or 10,000 ppm 8.6% citranaxanthin dry powder (equivalent to 86, 284, and 860 ppm active ingredient). Another group was administered dry powder without active ingredient. A control group remained untreated. In all groups with citranaxanthin the faeces had a remarkable red-brown colour. At the highest level the faeces were more liquid than normal. Feed intake decreased in a dose-dependent manner due to palatability problems. The dry powder group without citranaxanthin also had decreased feed intake.

Appearance, behaviour, weight gain, haematology, clinical chemistry, electrocardiography, urinalysis, weight of organs, and their autopsy did not reveal any influence of treatment. There were no pathological or histological changes in organs or tissues (Leuschner et al., 1975).

Observations in man

No information available.

COMMENTS AND EVALUATION

The major present-day use of citranaxanthin is as an animal feed additive to impart a yellow colour to chicken fat and egg yolk. It may also be used as a colouring agent by adding it directly to food.

If the substance were to be used as a direct food colouring agent, the data were not sufficiently comprehensive for evaluation (e.g. only one lifetime feeding study was available). The Committee concluded that further data of the type outlined in Annex III of "Principles for the Safety Assessement of Food Additives and Contaminants in Food" for synthetic food colours are required before the substance can be fully evaluated for direct food use (Annex 1, reference 76).

For its use as an animal feed additive, an evaluation could not be made because the data base did not include sufficient information on the nature of residues to be found in animal-derived foodstuffs and because there was no information concerning the use levels that would constitute good animal husbandry practice.

REFERENCES

BASF (1972a). Ergebnisse der gewerbetoxikologischen Vorprüfung. Unpublished report from BASF Department of Toxicology. Submitted to WHO by BASF Aktiengesellschaft, Ludwigshafen, FRG.

BASF (1972b). Report on tests for acute oral toxicity of citranaxanthin dry powder in rats. Unpublished report from BASF Department of Toxicology. Submitted to WHO by BASF Aktiengesellschaft, Ludwigshafen, FRG.

Crina (1974a). Citranaxanthin for broilers No. 1. Unpublished report. Submitted to WHO by BASF Aktiengesellschaft, Ludwigshafen, FRG.

Crina (1974b). Citranaxanthin for broilers No. 2. Unpublished report. Submitted to WHO by BASF Aktiengesellschaft, Ludwigshafen, FRG.

Hempel, K.J., Zeller, H., Kirsch, P., Koob, K., & Freisberg, K.O. (1973). Report on the examination of citranaxanthin 10% dry powder in a 91-day feeding test with rats. Unpublished report from Allgemeines Krankenhaus, Pathological Institute of Heidberg, Hamburg and BASF, Department of Toxicology. Submitted to WHO by BASF Aktiengesellschaft, Ludwigshafen, FRG.

Kawase, S., Komatsu, Y., Suzuki, Y., Nishida, S., & Kobayashi, A. (1972). Subacute toxicity of citranaxanthin. J. Med. Soc. Toho., Japan, 19, 499-504.

Kolk, J.H.H. (1974). About vitamin A efficiency and pigmenting effect of three citranaxanthin preparations at chicks and quails with different crystal size. Dissertation, University of Munich.

Leuschner, F., Leuschner, A., Schwerdtfeger, W., & Dontenwill, W. (1975). Oral toxicity of citranaxanthin, batch BASF No. 10 - called for short "cn" - and of the substance without agent - called for short "swa" - in the beagle dog (repeated dosage over 6 months). Unpublished report from Laboratorium für Pharmakologie und Toxikologie, Hamburg, to BASF. Submitted to WHO by BASF Aktiengesellschaft, Ludwigshafen, FRG.

Leuschner, F., Hübscher, F., & Dontenwill, W. (1976a). Oral toxicity of citranaxanthin-trockenpulver, batch BASF No. 10 - called for short "cn" - and of the substance without agent, batch 908 E 953 called "swa" - in the Sprague-Dawley rat (reproduction study covering three succeeding generations). Unpublished report from Laboratorium für Pharmakologie und Toxikologie, Hamburg, to BASF. Submitted to WHO by BASF Aktiengesellschaft, Ludwigshafen, FRG.

Leuschner, F., Leuschner, A., Schwerdtfeger, W., & Dontenwill, W. (1976b). Two-year toxicity testing of citranaxanthin dry powder, batch BASF No. 10 - called for short "cn" - and of the substance without agent, batch 908 E 953 called "swa" - in the Sprague-Dawley rat. Unpublished report from Laboratorium für Pharmakologie und Toxikologie, Hamburg, to BASF. Submitted to WHO by BASF Aktiengesellschaft, Ludwigshafen, FRG.

Leuschner, F. (1976c). Prüfung der akuten Toxizität von citranaxanthin. Charge BASF NR. 10- kurz "CN" genannt- an mischrassigen Hunden bei peroraler Verabreichung. Unpublished report from Laboratorium für Pharmakologie und Toxikologie, Hamburg, to BASF. Submitted to WHO by BASF Aktiengesellschaft, Ludwigshafen, FRG.

Morgenthaler, H. (1978). Citranaxanthin balance and excretion study. Unpublished report from BASF Pharma Division. Submitted to WHO by BASF Aktiengesellschaft, Ludwigshafen, FRG.

Morgenthaler, H. (1979). Organ citranaxanthin - distribution. Unpublished report from BASF Pharma Division. Submitted to WHO by BASF Aktiengesellschaft, Ludwigshafen, FRG.

Tiews, J. (1968a). Certified report about vitamin A activity of citranaxanthin and other carotenoids. Unpublished report. Submitted to WHO by BASF Aktiengesellschaft, Ludwigshafen, FRG.

Tiews, J. (1968b). Citranaxanthin for egg yolk pigmentation. Proceedings from Lohmann Nutrition Conference, Cuxhaven, pp. 47-50.

MISCELLANEOUS FOOD

ADDITIVES

L-GLUTAMIC ACID AND ITS AMMONIUM, CALCIUM, MONOSODIUM AND POTASSIUM SALTS

EXPLANATION

These substances were evaluated at the fourteenth and seventeenth meetings of the Committee (Annex 1, references 22 and 32). The ADI of 0-120 mg/kg b.w. allocated to L-glutamic acid encompassed the glutamic acid equivalents of the salts and was additional to glutamic acid intake from all non-additive dietary sources. The ADI did not apply to infants under 12 weeks of age and further work required if the use was to be extended to infant foods included the determination of oral non-adverse effect levels of glutamates in neonatal animals and age correlations between neonatal experimental animals and the human infant.

Since the last review additional data have become available and are summarized and discussed in the following monograph. The previously-published monographs have been expanded and are incorporated into this monograph.

BIOLOGICAL DATA

Biochemical aspects

Metabolism and pharmacokinetics

Glutamic acid is metabolized in the tissues by oxidative deamination (von Euler _et al._, 1938) or by transamination with pyruvate to yield oxaloacetic acid (Cohen, 1949) which, via alpha-ketoglutarate, enters the citric acid cycle (Meister, 1965). Quantitatively minor but physiologically important pathways of glutamate metabolism involve decarboxylation to γ-aminobutyrate (GABA) and amidation to glutamine (Meister, 1979). Decarboxylation to GABA is dependent on pyridoxal phosphate, a coenzyme of glutamic acid decarboxylase (Perrault & Dry, 1961), as is glutamate transaminase. Vitamin B_6-deficient rats have elevated serum glutamate levels and delayed glutamate clearance (Wen & Gershoff, 1972). A number of reviews on metabolism of glutamate that contain more comprehensive information have been published (Munro, 1979; Meister, 1979; Stegink, 1984).

Glutamate is absorbed from the gut by an active transport system specific for amino acids. This process is saturable, can be competitively inhibited, and is dependent on sodium ion concentration (Schultz et al., 1970). During intestinal absorption, a large proportion of glutamic acid is transaminated and consequently alanine levels in portal blood are elevated. If large amounts of glutamate are ingested, portal glutamate levels increase (Stegink, 1984). This elevation results in increased hepatic metabolism of glutamate, leading to release of glucose, lactate, glutamine, and other amino acids, into systemic circulation (Stegink, 1983c).

The pharmacokinetics of glutamate depend on whether it is free or incorporated into protein, and on the presence of other food components. Digestion of protein in the intestinal lumen and at the brush border produces a mixture of small peptides and amino acids; di- and tri-peptides may enter the absorptive cells where intracellular hydrolysis may occur, liberating further amino acids. Defects are known in both amino acid and peptide transport (Matthews, 1975, 1984).

Glutamic acid in dietary protein, together with endogenous protein secreted into the gut, is digested to free amino acids and small peptides, both of which are absorbed into mucosal cells where peptides are hydrolyzed to free amino acids and some of the glutamate is metabolized. Excess glutamate and other amino acids appear in portal blood. As a consequence of the rapid metabolism of glutamate in intestinal mucosal cells and in the liver, systemic plasma levels are low, even after ingestion of large amounts of dietary protein (Munro, 1979; Meister, 1979; Stegink, 1984).

Oral administration of pharmacologically high doses of glutamate results in elevated plasma levels. The peak plasma glutamate levels are both dose and concentration dependent (Stegink et al., 1973, 1974, 1975b, 1979a, 1982, 1983b, 1983c, 1985b; Ohara et al., 1977; Bizzi et al., 1977; Airoldi et al., 1979a; Daabees et al., 1985; Heywood et al., 1978). When the same dose (1 g/kg b.w.) of monosodium glutamate (MSG) was administered by gavage in aqueous solution to neonatal rats, increasing the concentration from 2% to 10% caused a five-fold increase in the plasma area under curve; similar results were observed in mice (Bizzi et al., 1977). Conversely, when MSG (1.5 g/kg b.w.) was administered to 43-day-old mice by gavage at varying concentrations of 2 to 20% w/v, no correlation could be established between plasma levels and concentration (James et al., 1978).

Administration of a standard dose of 1 g/kg b.w. MSG by gavage as a 10% w/v solution resulted in a marked increase of plasma glutamate in all species studied. Peak plasma glutamate levels were lowest in adult monkeys (6 times fasting levels) and highest in mice (12-35 times fasting levels). Age-related differences between neonates and adults were observed; in mice and rats, peak plasma levels and area under curve were higher in infants than in adults while in guinea pigs the converse was observed (Stegink et al., 1979a; Airoldi et al., 1979a; Ohara et al., 1977).

Studies on the effects of food on glutamate absorption have been carried out in mice, pigs, and monkeys. When infant mice were given MSG with infant formula or when adults were given MSG with consomme by gastric intubation, peak plasma glutamate levels were markedly lower than when the same dose was given in water, and the time to reach peak levels was longer (Ohara et al., 1977). The simultaneous administration of metabolizable carbohydrate was found to increase glutamate metabolism in mice, pigs, and monkeys, leading to lowered peak plasma levels (Stegink et al., 1979a, 1983a, 1983b, 1985a; Daabees et al., 1984). In contrast to gastric intubation, ad lib feeding of MSG in the diet caused only slight elevation of plasma glutamate above basal levels (Ohara et al., 1977; Airoldi et al., 1979a; Heywood et al., 1977; Yonetani & Matsuzawa, 1978).

Metabolic studies in humans

Similar effects of food on glutamate absorption and plasma levels have been observed in man. Only slight rises in plasma glutamate followed ingestion of a dose of 150 mg MSG/kg b.w. to adults with a meal; human infants, including premature babies, have the capacity to metabolize similar doses given in infant formula (Tung & Tung, 1980). Human plasma glutamate levels were much lower when large doses of MSG were ingested with meals compared to ingestion in water; in studies in which MSG was given with tomato juice, sloppy joes, Sustagen, Polycose, starch, or sucrose, metabolizable carbohydrate significantly lowered peak plasma glutamate levels (Bizzi et al., 1977; Byun, 1980; Ghezzi et al., 1980, 1985; Marrs et al., 1978; Stegink et al., 1979a, 1979b, 1982, 1983a, 1983b, 1983c, 1985a, 1986).

In general, foods providing metabolizable carbohydrate significantly attenuate peak plasma glutamate levels at doses up to 150 mg MSG/kg b.w. Carbohydrate provides pyruvate as a substrate for transamination with glutamate in mucosal cells so that more alanine is formed and less glutamate reaches the portal circulation (Stegink et al., 1983b).

Special studies on transplacental transport

When MSG (8 g/kg b.w.) was administered orally to rats on day 19 of gestation, maternal plasma levels rose from approximately 100 µg/ml to 1650 µg/ml, but no significant changes were observed in plasma glutamic acid of the fetuses (Ohara et al., 1970).

Infusion of MSG into pregnant rhesus monkeys at a rate of 1 g/hr led to a 10-20-fold increase in maternal plasma glutamate, but fetal levels remained unchanged. Higher rates of infusion resulted in maternal plasma glutamate levels up to 70 times basal levels, but fetal levels increased less than 10 times (Stegink et al., 1975a; Pitkin et al., 1979).

In vitro perfusion studies using human placenta indicated that the placenta served as an effective metabolic barrier to the transfer of glutamic acid (Schneider et al., 1979).

Special studies on the blood brain barrier

Glutamate levels are far higher in the brain than in plasma in mice, rats, guinea pigs, and rabbits (Garattini, 1971; Giacometti, 1979; Bizzi et al., 1977).

Efflux of glutamate from the brain has been reported to be seven times greater than influx, reflecting biosynthesis in the brain. The transport rate of glutamate from blood to the brain is much lower than for neutral or basic amino acids (Oldendorf, 1971). Normal plasma glutamate levels are nearly 4 times the K_m of the transport rate to the brain, so that glutamate transport systems are virtually saturated under physiological conditions (Pardridge, 1979).

In guinea pigs, rats, and mice, brain glutamic acid levels remained unchanged after administration of large oral doses of MSG which resulted in plasma levels increasing up to 18-fold (Peng et al., 1973; Liebschultz et al., 1977; Caccia et al., 1982; Airoldi et al., 1979a; Bizzi et al., 1977). Brain glutamate increased significantly only when plasma levels were about 20 times basal values following an oral dose of 2 g MSG/kg b.w. (Bizzi et al., 1977).

The failure to observe changes in whole brain glutamate when plasma levels are elevated does not preclude the possibility that levels in small regions, such as the arcuate nucleus of the hypothalamus, may increase (Perez & Olney, 1972).

Subcutaneous injection of high doses (2 g MSG/kg b.w.) to neonatal mice caused an increase in serum glutamate to 270 times basal values, while levels in the arcuate nucleus increased 4-7 fold (Price et al., 1981).

No appreciable changes in glutamate concentrations were observed in the lateral thalamus and in the arcuate nucleus of adult or neonatal rats given 4 g MSG/kg b.w. or 2 g MSG/kg b.w., respectively, by gavage. Peak plasma glutamate levels were 11-12 times normal after these doses (Airoldi et al., 1979b).

Endocrinology studies

Numerous studies have been carried out in which large doses of MSG were administered by subcutaneous or intraperitoneal injection to neonatal mice. A common effect of this treatment is a metabolic obesity without hyperphagia and stunted growth (Araujo & Mayer, 1973; Matsuyama et al., 1973; Nagasawa et al., 1974). The observed obesity in these studies was associated with decreased adrenaline-stimulated lipolysis. Decreased pituitary weight and impaired pituitary function resulted in atrophy of related target organs such as the gonads, accessory sexual organs, thyroids, and adrenals. Prolactin and growth hormone levels were depressed, but hypothalamic LHRF was reported to be unaffected (Lechan et al., 1976). Repressed ossification reported in one study was thought to be due to deranged PTH/calcitonin regulation (Dhindsa et al., 1978).

Similar experiments in rats also resulted in stunting and obesity, with reduction in weights of the pituitary, adrenals, and gonads. Growth hormone levels were reduced in both sexes but LHRH, TRH, somatostatin, and norepinephrin levels were unaffected. Rats receiving 1 g MSG/kg b.w. subcutaneously showed elevated prolactin but reduced growth hormone and TSH levels (Nemeroff et al., 1975, 1977a,b,c; Redding et al., 1971).

The effects of treatment are age-dependent in both mice and rats. Neonatal rats show a permanent reduction in GH secretion without evidence of excessive prolactin secretion whereas acute administration of MSG to adults causes suppression of GH and PRL release by effects on the dopamine systems in the medial basal hypothalamus (Terry et al., 1977). Reduction in weight of the endocrine glands without obvious histological changes did not affect fertility (Trentini et al., 1974; Lengvari, 1977).

Physiological role of MSG

L-Glutamate and GABA supposedly act as excitatory and inhibitory transmitters, respectively, in the central nervous system. Glutamate is also involved in the synthesis of proteins (Krnjevic, 1970).

Taste physiology

Chemical senses (taste and olfaction) affect the cephalic phase of secretion of gastric acid, the exocrine pancreas, gastrin, glucagon, insulin, and pancreatic polypeptide hormone (Brand et al., 1982).

MSG has a unique taste, called umami (Ikeda, 1908, 1909, 1912) and in addition glutamate has flavour-enhancing properties in some foods (Kirimura et al., 1969; Solms, 1969; Yamaguchi, 1979). A detailed review of glutamate and the umami taste has been published and deals with physiological and psychological aspects (Kawamura & Kare, 1987).

Nutritional aspects

L-Glutamic acid occurs as a common constituent of proteins and protein hydrolysates. Nutritional studies in the rat have shown glutamic acid to be a non-essential amino acid that is required in substantial amounts to ensure high growth rates in rats (Hepburn et al., 1960). Some interconversion between glutamic acid and arginine can occur to cover minor dietary deficiencies (Hepburn & Bradley, 1964).

As an essential substrate in intermediary metabolism, glutamate is present in organs/tissues in the concentrations shown in Table 1 (Giacometti, 1979).

The free amino acid pools in the tissues constitute about 70 g in the adult, of which the major components are alanine, glutamic acid, glutamine, and glycine. The daily turnover of glutamic acid in a 70 kg man has been estimated as 4,800 mg (Munro, 1979). Human plasma contains 4.4-4.5 mg/l of free glutamic acid and 9 mg/l bound glutamic acid; human urine contains 2.1-3.9 µg/mg creatinine of free glutamic acid and 200 µg/mg creatinine of bound glutamic acid (Peters et al., 1969). Human spinal fluid contains 0.34-1.64 (mean 1,03) mg/l free glutamic acid (Dickinson & Hamilton, 1966).

Table 1. Free glutamate concentrations in organs and tissues

Organ/tissue	Total Free Glutamate (mg)
Muscle	6,000
Brain	2,250
Kidneys	680
Liver	670
Blood plasma	40
Total	9,640

Premature and full-term infants hydrolyze any given protein in the stomach to very similar extents (Berfenstam et al., 1955). Hepatic glutamate dehydrogenase appears at 12 weeks of human fetal life; it is present in rat fetal liver on day 17 and reaches its maximum within 2 weeks after birth (Francesconi & Villee, 1968). Gouty patients have raised levels of plasma glutamate compared to normal and, following a protein meal, glutamate reaches excessive levels (Pagliari & Goodman, 1969).

Glutamine and glutamic acid are the most abundant amino acids in the milk of all species; human milk contains 1.2% protein, of which 20% is bound glutamic acid, equivalent to 3 g/l calculated as MSG. The free glutamic acid concentration is about 300 mg/l. In contrast, cow's milk contains 3.5% protein but only 30 mg/l free glutamic acid (Maeda et al., 1958, 1961). Later studies have indicated that human milk contains about 600 μmole glutamate/l from days 1-7 post-partum, rising to 1,300-1,500 μmole/l thereafter. Human or chimpanzee milk is 10 times higher in free glutamate than is rodent milk (Rassin et al., 1978).

Daily intake of free glutamic acid by the breast-fed infant has been estimated to be about 36 mg/kg b.w. (equivalent to 46 mg/kg b.w. as MSG) while daily intake of protein-bound glutamate was estimated as approximately 360 mg/kg b.w. The breast-fed infant in the USA ingests more glutamate, on a body-weight basis, than at any other time of life (Baker et al., 1979).

High levels of free glutamic acid have been found in cantaloupe (0.5 g/kg) and grapes (0.4 g/kg), while high levels of aspartic acid were found in figs (2.6 g/kg), nectarines (2.0 g/kg), peaches (1.1 g/kg), yellow plums (1.8 g/kg), and dry prunes (1.9-5.2 g/kg) (Fernandez-Flores et al., 1970). Fish and meat had less than 0.1 g/kg of free glutamic acid, sausage 0.1-1.5 g/kg, cheese 0.2-22 g/kg, "tomatenflocken" 15 g/kg, and dried mushrooms 17 g/kg (Mueller, 1970).

Daily intakes of free and bound glutamate by breast-fed infants at 3 days of age were 1.10 g bound and 0.115 g free, corresponding to 0.408 g/kg b.w. At one month of age, intakes were 1.37 g bound and 0.144 g free, corresponding to 0.405 g/kg b.w. Infants aged 5-6 months receiving 500 g cow's milk and 2 jars baby food per day would have a daily intake of 4.0 g bound and 0.075 g free glutamic acid, equivalent to 0.62 g/kg b.w. (Berry, 1970). The mean daily intake of MSG of individuals over 2 years of age has been estimated as 100-225 mg per capita, an increment of 3-7% over the glutamic acid supplied by dietary protein (GRAS, 1976).

Consumption of MSG in various countries has been estimated (see Table 2) (Maga, 1983).

Table 2. MSG consumption in several countries

Country	MSG Consumption (g/day)
Taiwan	3.0
Korea	2.3
Japan	1.6
Italy	0.4
USA	0.35

Male weanling rats were fed casein or purified amino acid diets for 14 or 21 days. The addition of 0.33% glycine or 1.14% glutamic acid to a diet with a protein equivalent (N x 6.25) of 10% essential amino acids increased food efficiency (g weight gain/g food) from 0.38 ± 0.01 to 0.41 ± 0.01 (Adkins et al., 1967).

The nutritional value of non-essential amino acids as the nitrogen source in a crystalline amino acid diet for chick growth was examined. The basal diet contained 27% of an essential amino acid mixture and an additional 24% of non-essential amino acids. In the test diets all non-essential amino acids were removed, and single amino acids were added as follows: glutamic acid 11.3%, aspartic acid 10.2%, alanine, 6.8%, glycine 5.8%, proline 8.7% (additional to the 1% in the basal diet), or serine 8.1%. Of these, glutamic acid and aspartic acid were found to be very useful nitrogen sources, alanine was useful, glycine and proline were insufficient, and serine was harmful for chick growth. The chicks fed the L-glutamic acid diet showed less growth than those fed the basal diet, although the differences between the groups were not statistically significant (Sugahara & Ariyoshi, 1967).

In an 80-day feeding study, young rats received a diet containing 8% milk protein and 0, 2, 4, or 6% MSG. Body-weight gain and body-nitrogen content were significantly greater in the groups receiving 4 or 6% MSG relative to controls. No hypothalamic pathology was observed. When MSG was added to the diet of rats containing 12% milk protein at levels of 2 or 8% for six weeks, an effect on body-weight gain was observed (Huang et al., 1976).

Toxicological studies

Special studies on mutagenicity

Cells (kangaroo rat cell line) were exposed continuously for 72 hours to 0.1% monosodium glutamate without showing any toxic effects (US FDA, 1969).

Groups of 12 male albino Charles-River mice received single oral doses, by gavage, of monosodium glutamate at levels of 0, 2.7, or 5.4 g/kg b.w. The treated animals were mated with groups of 3 untreated females for each of 6 consecutive weeks. Females were sacrificed at mid-term of pregnancy, and the uteri were examined for signs of early embryonic death. Females that had been mated with treated males showed no differences compared to controls in the number of implantations, resorptions, and embryos (Industrial Bio-Test, 1973a).

In a host-mediated assay using Salmonella typhimurium G46, the bacteria were administered intraperitoneally to the host rats which received 0.2 or 5.7 g MSG/kg b.w. orally for 14 days. No increase in revertants was seen relative to controls (Industrial Bio-Test, 1973b).

Potassium and ammonium glutamate, L-glutamic acid, and L-glutamic acid·HCl were not mutagenic when tested against S. typhimurium strains TA98, TA100, TA1535, TA1537, and TA1538 and Saccharomyces cereviseae in the presence or absence of S-9 mix (Litton Bionetics, 1975a, 1975b, 1977a, 1977b).

Special studies on neurotoxicity

Mice and rats

The effects of glutamate on cerebral metabolism were studied by intraventricular injection of L-glutamic acid into mice; 150 mg produced convulsions, uncoordinated grooming, or circling of the cage (Crawford, 1963).

Two percent intraarterial sodium glutamate increased epileptic fits and intracisternal L-glutamic acid caused tonic-clonic convulsions in animals and man. High parenteral doses of L-glutamic acid caused EEG changes only in dogs with previous cerebral damage, and no rise was detected in the cerebral spinal fluid level of glutamate (Herbst et al., 1966).

Monosodium glutamate injected i.p. at a level of 3.2 g/kg b.w. caused reversible blockage of beta waves in the electroretinogram in immature mice and rats, indicating retinotoxicity (Potts et al., 1960). The timing of treatment of mice was quite critical. After 10-11 days postnatal it was difficult to produce significant lesions of the retina (Olney, 1969a). A study of the glutamate metabolizing enzymes of the retina of glutamate-treated rats indicated a decrease in glutaminase activity, an increase in glutamic aspartate transaminase, and no change in glutamyl synthetase and glutaminotransferase. The effects appear to be due to repression and induction of enzyme synthesis. Glutamate uptake by the retina, brain, and plasma decreases with age and is slower in 12-day old animals than in 5-day old animals (Freedman & Potts, 1962, 1963).

Subcutaneous injection of L-monosodium glutamate at 4-8 g/kg into mice caused retinal damage with ganglion cell necrosis within a few hours. In very young animals there was extensive damage to the inner layers (Lucas & Newhouse, 1957).

Neonatal mice aged 9-10 days were given single subcutaneous injections of 4 g/kg monosodium glutamate; the animals were killed from 30 minutes to 48 hours later. The retinas showed acute lesions on electron microscopy with swelling dendrites and early neuronal changes leading to necrosis followed by phagocytosis (Olney, 1969a).

Mice aged 2 to 9 days were killed 1-48 hours after single subcutaneous injections of 0.5-4 g/kg monosodium glutamate. Lesions were seen in the preoptic and arcuate nuclei of the hypothalamic region on the roof and floor of the third ventricle and in scattered neurons in the nuclei tubercles. No pituitary lesions were seen, but subcommissural and subfornical organs exhibited intracellular oedema and neuronal necrosis. Adult mice given subcutaneously 5-7 g/kg monosodium L-glutamate showed similar lesions (Olney, 1969b).

Degeneration of neonatal mouse retina has been reported following parenteral administration of MSG (10 subcutaneous injections of 2.2-4.2 g/kg 1-10 days after birth) (Cohen, 1967).

Sixty-five neonatal mice aged 10-12 days received single oral doses of monosodium glutamate at 0.5, 0.75, 1.0, or 2.0 g/kg b.w. by gavage. Ten were controls and 54 mice received other amounts. After 3 to 6 hours all treated animals were killed by perfusion. Brain damage, as evidenced by necrotic neurons, was evident in arcuate nuclei

of 51 animals: 52% at 0.5 g/kg, 81% at 0.75 g/kg, 100% at 1 g/kg, and
100% at 2 g/kg. The lesions were identical by both light and electron
microscopy to subcutaneous-produced lesions.   No lesions were seen at
0.25 g/kg. The number of necrotic neurons rose approximately with
dose. Four animals tested with glutamic acid also developed the same
lesions at 1 g/kg b.w. The effect was additive with aspartate (Olney,
1970).

High subcutaneous doses of 1 or 4 g/kg of MSG caused hypothalamic changes in 42 and 60%, respectively, of treated 5 to 7-day old
mice. Oral administration (1 or 4 g/kg) of a 4% aqueous solution
elicited a predominantly glial reaction in 26-28% of the mice. The
remainder were unaffected (Abraham et al., 1971).

Six 9 to 10-day old mice, dosed orally with 10% monosodium
glutamate (2 g/kg), showed characteristic brain lesions (Geil, 1970).

Monosodium   glutamate,   monopotassium   glutamate,   sodium
chloride, and sodium gluconate at 1 g/kg in a 10% w/v solution (and
comparable volumes of distilled water), were administered orally and
subcutaneously to mice and rats at 3 or 12 days of age and to dogs at 3
or 35 days of age and the animals were killed within 24 hours of dosage.
Examination of the eyes and of the preoptic and arcuate nuclei of the
hypothalamus by two pathologists revealed no dose-related histomorpho-
logical effects in any of the test groups at either of the two ages
selected to correspond to the periods before and at the beginning of
solid food intake (Oser et al., 1971).

Seventy-five infant Swiss albino CD-1 mice (3 to 10 days old)
were given single subcutaneous injections of MSG at concentrations
equal to 2 or 4 g/kg (0.1 ml in distilled water). Another group of
50 adult CD-1 mice were injected either subcutaneously or intraperitoneally with MSG at doses varying from 6 to 10 g/kg (1 ml volume).
Control animals were injected with sodium chloride. Brain tissue was
examined by light and electron microscopy. Ninety-five percent of the
animals injected with MSG developed brain lesions in the arcuate nucleus
of the hypothalamus. Lesions involved primarily microglial cells, with

no effects to the perikarya of neurons. Distal neuronal processes were
only slightly affected (Arees & Mayer, 1971).

Thirteen neonate CF1-JCL mice received single subcutaneous
injections of 1 g/kg of MSG at days 2 and 4 after birth. Brains were
removed 1, 3, 6, and 24 hours post-injection and examined by light
microscopy. The common finding after 3 and 6 hours was necrosis of the
neural element in the region of the hippocampus and hypothalamus. When
pregnant mice of the same strain were injected subcutaneously with
5 mg/g on days 17 and 18 of pregnancy, examination of fetal brains 3,
6, and 24 hours after treatment showed cellular necrosis in both the
ventromedial and arcuate nuclei (Murakami & Inouye, 1971).

Groups of 3- and 12-day old C57BL/J6 mice, each containing
5 animals, were given single intragastric or single subcutaneous doses
of monosodium glutamate, sodium chloride, sodium gluconate, potassium
glutamate (all 10% solutions, 10 ml/kg b.w.), or water. All animals
were killed 24 hours after dosing. Microscopic examination of the
brains, particularly the ventral hypothalamus, did not show any neuronal
necrosis of the hypothalamic arcuate nuclei (Oser et al., 1973).

Groups of 10-day old Swiss Webster albino mice, each
containing 10 animals, were given single subcutaneous doses of one of
24 compounds structurally related to MSG. The dose level was either 12
or 24 μmole/kg b.w. Five hours post-dosing the brains and retinas
were processed for light or electron microscopy. Roughly quantifying
the pathological reaction in the infant hypothalamus was used as a
method for comparing the neurotoxic potency of the test compounds.
Except for L-cysteine, all neurotoxic compounds were acidic amino acids
known to excite neurons. The most potent neurotoxic compounds were
those known to be powerful neuroexcitants (N-methyl-DL-aspartic and
DL-homocysteic acids) (Olney et al., 1971).

Groups of 10-12-day old Swiss Webster albino mice, each
containing 7-23 animals, were given single oral doses of MSG at levels
of 0.25, 0.50, 0.75, 1.0, or 2.0 g/kg. Groups of 2 or 4 mice of the same

age were given single oral doses of either 1.0 or 3.0 g/kg L-glutamic acid or monosodium-L-aspartate or 3.0 g/kg L-glutamate-L-aspartate, monosodium glutamate, NaCl, L-glycine, L-serine, L-alanine, L-leucine, D,L-methionine, L-phenylalanine, L-proline, L-lysine, L-arginine, or L-cysteine. The animals were sacrificed after dosing and brains were examined by either light or electron microscopy. The severity of brain damage was estimated by quantifying the pathological changes in the hypothalamus. One g/kg of glutamic acid destroyed approximately the same number of hypothalamic neurons as a comparable dose of MSG. Of the amino acids tested, only aspartate and cysteine produced hypothalamic damage. These amino acids caused both retinal and hypothalamic lesions similar to those found after treatment with MSG (Olney & Ho, 1970).

Infant litter mates of Swiss Webster mice were divided into two groups. The experimental group received single daily subcutaneous injections of MSG for 10 consecutive days. The control litter-mate group received injections of 0.9% saline. All injections were of 0.02 ml volume. The dose of the first MSG injection, starting 24 hours after birth, was 2 g/kg b.w. Subsequent daily doses were increased by 0.25 g/kg per day so that the final dose on the tenth day was 4.25 g/kg. When the surviving MSG-treated and control mice attained 20-28 g b.w., they were subjected to a battery of behavioural and pharmacological tests. The study period lasted until 50 days after birth. There were no significant or observable differences in response to behavioural tests or to selected drugs (Prabhu & Oester, 1971).

A group of 10-day old mice were given single subcutaneous doses of 18 mmole MSG/kg b.w. and sacrificed from 15 minutes to 8 days subsequently. Further groups were given 2 g, rising to 4 g MSG/kg b.w. daily, from day 1 to day 10 and sacrificed 9 months later, or 4 g MSG/kg b.w. by gavage on day 10. There was quite irreversible damage to neurons in the arcuate nucleus and rapid cell necrosis; there were fewer cells 2-4 days later. An early, reversible glial and ependymal oedema was also seen (Olney, 1971).

Neonatal mice were given 0, 2, or 4 g MSG/kg b.w. by gavage and sacrificed after 20 or 30 minutes or after 1, 2, 3, or 24 hours. At the higher-dose level, oedema and necrosis of the neurons of the arcuate nucleus were seen after 20 minutes and preoptic and arcuate nucleus lesions were observed after 30 minutes. The lesions spread wide with time affecting the tectum and other structures in 2-3 hours. Phagocytosis was seen in the arcuate nucleus after 24 hours. Primary lesions in neurons were seen at the electron microscope level after 30 minutes (Lemkey-Johnston & Reynolds, 1972, 1974).

Sodium chloride was administered to 3-9-day old mice as a 10% solution by gavage at doses of sodium equivalent to 1-10 g MSG/kg b.w.; glutamic acid·HCl was administered at 2 and 4 g/kg b.w. as a 20% solution and sucrose (80% w/v) was given at dosages equivalent to 4, 8, or 10 g MSG/kg b.w. Oedema and pyknotic nuclei were seen in sodium chloride-treated animals of 5 days of age; no lesions were seen in animals older than 6 days at the time of treatment, nor in mice given sucrose (Lemkey-Johnston et al., 1975).

Forty male and 41 female mice were given daily subcutaneous injections of 2.5 g MSG/kg b.w. from 5-10 days of age and were subsequently reared to maturity. Adult animals had an 80% decrease in perikarya of the arcuate nucleus, endocrine deficits, reduced reproductive performance, stunted growth, obesity, and decreased weights of the pituitary, ovaries, and testes (Holzwarth-McBride et al., 1976).

Monosodium glutamate was administered to 4 weanling mice of each sex ad libitum in the diet or drinking water at levels of 46 g/kg/day or 21 g/kg/day, respectively. No hypothalamic lesions were induced. Plasma glutamic acid levels were doubled by giving MSG at 10% w/v in the diet, but the threshold for neurotoxicity of MSG by dietary administration was not exceeded (Heywood et al., 1977).

Groups of 24 mice were given single subcutaneous injections of 1, 2, 3, or 4 g MSG/kg b.w. at age 10 days, or 2, 4, or 6 g MSG/kg b.w. at age 60 days. Examination of the brains 3-5 hours after treatment showed damage to the area postrema (Olney et al., 1977).

Weanling mice were fasted and deprived of fluids overnight, then given as drinking fluid either 10% MSG; a mixture of 5% MSG, 4.5% monosodium aspartate and 1% aspartame; 5% MSG; 2.5% MSG and 2.5% aspartate; or a choice between 10% MSG and deionized water. The brains were examined about 4 hours after start of treatment. All animals ingesting MSG-containing solutions sustained hypothalamic injury (Olney et al., 1980).

Neonatal mice aged 6-12 days were injected with single doses of 0.25, 0.5, 1, 2, or 4 g MSG/kg b.w. Hypothalamic lesions were evident at doses of > 0.5 g/kg b.w. (Reynolds et al., 1976).

A total of 52 mice, 10 days of age, were given single intraperitoneal doses of 4 g MSG/kg b.w. Two mice were killed every hour from 1 to 20, then at 24, 48, 72, 96, and 168 hours after injection and the brains were examined. Neuronal changes appeared 2 hours after dosing, comprising somatic ballooning, nuclear chromatin clumping, pyknosis, and cytoplasmic vacuolation in the neurons of the arcuate nucleus. The histopathologic changes became more severe with time (Takasaki, 1978a).

A total of 38 mice, 2-20 days old, were given single intraperitoneal doses of 4 g MSG/kg b.w.; the animals were killed 8 hours after injection and the brains were examined. All treated animals showed necrotic changes, including pyknosis and karyorrhexis, of neurons of the arcuate nucleus; the number of necrotic neurons was maximal in mice dosed at 6-8 days of age and diminished in severity with age of dosing until 20 days of age, when few necrotic neurons were seen (Takasaki, 1978a).

Four mice, 10 days old, were given single intraperitoneal doses of 4 g MSG/kg b.w., killed 8 hours after treatment, and the brains examined to determine the total extent of the lesions. Acute necrotic changes were noted in the subfornical organ, preoptic area, and area postrema, in addition to the arcuate nucleus; in 1 mouse pyknotic neurons were also found in the cerebral cortex (Takasaki, 1978a).

Eighteen 1-day old mice were injected intraperitoneally daily with doses of 4 g MSG/kg b.w. for 1-10 days and killed 8 hours after the last dose. No neurons disappeared after a single injection but neurons that withstood a single injection were subsequently affected and disappeared almost completely by the fourth consecutive daily injection (Takasaki, 1978a).

Twenty-two mice, 10 days old, were injected with 0.1-2 g MSG/kg b.w. intraperitoneally, killed 8 hours later, and the hypothalamic arcuate region examined. A further 120 mice received doses of 0.1-4.0 g MSG/kg b.w. by gavage and were examined similarly. The threshold dose was 0.4 g MSG/kg b.w. i.p. or 0.7 g/kg b.w. by gavage (Takasaki, 1978a).

Pregnant mice received MSG at levels of 0, 5, 10, or 15% in the diet or 5% in drinking water on the eighteenth day of gestation. The animals were killed and fetal brains were examined. Groups of lactating mice were put on the same regimes for 1-4 days and the brains of the pups examined. Further groups of weanling mice were given similar diets for 1-4 days and killed 3 hours after feeding. No changes were seen in the arcuate nucleus of the fetuses, suckling pups, or weanlings (Takasaki, 1978b).

The smallest neurotoxic dose increases with age and oral doses give lower plasma levels than parenterally-administered doses. Dietary feeding of MSG does not affect serum levels of LH or testosterone. Small subcutaneous doses given to neonatal and infant mice or large oral doses given to weanling mice produced no adverse effects (Takasaki et al., 1979a).

Groups of 10-day old mice were given by gavage 2 or 4 g MSG/kg b.w. together with 0.63 g NaCl/kg b.w. or 1.93 g glucose/kg b.w. Sodium chloride did not potentiate MSG-induced brain lesions, while glucose reduced significantly the number of neurotic neurons in

the arcuate nucleus. In another experiment, groups of 10-day old mice
were given by gavage 2 g MSG/kg b.w. alone or together with 1.93 g/kg
b.w. of glucose, fructose, galactose, or lactose, or 1.83 g/kg b.w. of
sucrose. The mono- and disaccharides reduced significantly the number
of neurons affected in the arcuate nucleus (Takasaki, 1979).

Groups of infant mice, 10 days of age, were given by gavage
2 g MSG/kg b.w. alone or 2 g MSG/kg b.w. along with 2.3 g arginine•HCl/
kg b.w., 0.2 g leucine/kg b.w., or 0.02 units of insulin (prior to
gavage). All the additional treatments reduced the extent of injury to
the arcuate nucleus relative to MSG alone (Takasaki & Yugari, 1980).

Weanling mice were deprived of water for 14 hours overnight
and then given solutions of MSG as the sole drinking fluid; 12-
180 animals receiving > 3 g MSG/kg b.w. in drinking water developed
lesions of the arcuate nucleus. Fewer lesions were observed if free
choice was allowed between water and MSG solutions, if glucose or
arginine were also added to the MSG solutions, or if water deprivation
occurred during the day (Torii & Takasaki, 1983).

Suckling mice were given by gavage 2, 6, or 9 mg MSG/kg b.w.
on days 6-10 postnatally and were killed on day 11. No hypothalamic
lesions were noted. Another group given 6 mg MSG/kg b.w. for 5 days
were weaned and observed for 12 months. No hypophagia, obesity, or
hyperactivity were noted (Wen et al., 1973).

Casein and fibrin hydrolysates were administered to 9-11-day
old mice at dose levels of 20, 50, or 100 µl/g b.w. and plasma amino
acid levels were determined at intervals. No neuronal damage resulted
when plasma glutamate levels were below 24 µmole/dl (normal, 6-10 µmole/
dl); the minimal threshold plasma level for neuronal injury was
estimated as about 50-52 µmole/dl. Older (25 days old) mice showed a
markedly greater ability to metabolize glutamate, which probably
accounts for their decreased sensitivity to glutamate (Stegink et al.,
1974).

MSG was administered at dose levels of 1 g/kg to infant mice and 2 and 4 g/kg to infant rats. All the animals developed lesions in the arcuate area of the hypothalamus and median eminence. No evidence of cellular pathology was noted in controls (Burde et al., 1971).

Rats

Weanling Sprague-Dawley male rats were given 20 mmoles MSG intraperitoneally (equal to 3.4 g/kg b.w.). Marked somnolence was observed within 5 to 20 minutes. About one-third to one-half of the animals salivated copiously and had myocolonic jerking about 1 hour after injection, sometimes followed by vigorous running about the cage and stereotyped biting (Bhagavan et al., 1971).

At birth, the rat brain glutamate concentration is about 4 mM and increases over a period of 20 days to the adult value of approximately 10 mM. When a 4 g/kg dose was given intragastrically, convulsions were seldom observed, and then only after 90 minutes. Two g/kg MSG given intraperitoneally always caused convulsions. When young rats were given 4 g/kg MSG, monosodium aspartate, or glycine, the glutamine level was increased significantly in the brain in all cases, but only monosodium glutamate and aspartate caused convulsions. D-Glutamate (4 g/kg), which is not deaminated by the rat, also caused convulsions. These results suggest that the convulsions caused by MSG are not due to liberated ammonia, but rather to the amino acid anion. At 4 g/kg, MSG gave rise to serum concentrations of glutamate of about 70 mM, strongly suggesting osmotic problems (Mushahwar & Koeppe, 1971).

An experiment was conducted in which 45 male and 45 female rats were fed 1 or 250 mg/kg MSG daily from 1 to 90 days of age, at which time the animals were killed. A comparable control group received only laboratory chow over the same period of time. General clinical observations, body weights, haematologic parameters, and other clinical chemical measurements were within the normal range. At autopsy, organ weights were within the normal range. Histochemical and ultrastructural studies of the hypothalamus and median eminence showed no evidence of repair or replacement of neuronal cells by elements of glial or ependymal cells (Golberg, 1973).

Groups of 10-day old Charles-River rats (10 male and 10 female) were dosed orally with 0.2 ml of either strained baby food containing no monosodium glutamate, strained baby food containing monosodium glutamate up to 0.4%, or strained baby food containing monosodium glutamate equal to a dosage level of 0.5 g/kg, additional to that found in normal commercially distributed baby food (390 mg per jar). The rats were mated; half of the offspring were removed from parental females and sacrificed after 5 hours. Histological studies were made of brains in the area of the hypothalamus at the roof and the floor of the third ventricle. The remaining rats were returned to parental females and allowed to grow to maturity (90 days post-weaning), then sacrificed and histological studies made of the brain. No lesions were observed in the brains of animals sacrificed at either 5 hours post-treatment or after reaching maturity. Animals which were reared to maturity showed normal growth and food consumption (Geil, 1970).

Male and female Wistar rats, 3 to 4 days old, received single subcutaneous injections of MSG. The total dose was 4 g/kg b.w. in a volume of 0.1 ml. Control rats were injected with an equal volume of saline. In experiments concerned with the acute effects of MSG on the brain, the infant rats were killed 3 hours after the injection of MSG. Light microscopic and electron microscopic examinations failed to reveal any effects upon the lateral preoptic nucleus, arcuate nucleus, or median eminence. To determine long-range effects of MSG, uniform litters (8 pups per mother) were kept in an environmentally-controlled room until weaned. The experiment was terminated at 68 days post-treatment for males and 88 days post-treatment for females. In the MSG-treated females, the relative ovarian weights were significantly less than in the controls (29.3 ± 1.4 mg/100 g versus 34.7 ± 0.9 ($P < 0.1$)), but otherwise there were no significant differences in the weight of reproductive organs (ovaries, testes, seminal vesicles, prostate). The adult MSG-treated females cycled normally and were capable of mating and producing normal litters (Adamo & Ratner, 1970). In a letter to Science, it is pointed out that Adamo & Ratner injected a 40% solution of MSG instead of the 10% solution used by Olney et al. (1971).

Fifty weanling rats (Food and Drug Research Labs strain), 3 days of age, were divided into five groups. Each group was subdivided and the test animals given either single intragastric or single subcutaneous doses of monosodium glutamate, sodium chloride, sodium gluconate, potassium glutamate (all 10% solutions, 10 mg/kg b.w.), or water. Another group of 12-day old rats were treated in a similar manner. All animals in these groups were killed 24 hours after dosing. In addition, another group of 60 rats (3 days of age) was divided into subgroups and treated with the same test compounds at the same dose levels. Half of each group was sacrificed at 6 hours and the other half at 24 hours after dosing. Microscopic examination of the brain, particularly the ventral hypothalamus, did not show neuronal necrosis of the hypothalamic arcuate nuclei, except in 1 rat dosed with 1 g/kg monosodium glutamate at 3 days of age, and killed 24 hours later, which showed an area in the median eminence which contained cells with slight nuclear pyknosis and prominent vacuolation (Oser et al., 1973).

Neonatal rats were given MSG by injection on days 1-14 and the retina, optic nerve, and optic tracts were examined 2-10 months later. Observed changes included destruction of the ganglion cell layer and the inner nuclear layer together with a reduced number of myelinated axons (Hansson, 1970).

Rats were given by injection 0.25-4 g MSG/kg b.w. from the sixteenth day of pregnancy (in utero exposure) to the twenty-first day post partum. The animals were killed 1, 2, or 3 hours after injection. Lesions were seen in the cerebral cortex, hippocampus, thalamus, and hypothalamus. The earliest changes were lysis of organelles followed by oedema of the parenchyma and finally neuronal pyknosis (Everly, 1971).

Male rats aged 23 days were deprived of water or water and food overnight, and then given aqueous solutions of MSG at concentrations up to 10%. No hypothalamic lesions were found with intakes up to 5 g MSG/kg b.w. (Takasaki & Torii, 1983).

Five groups of 6 weanling rats were given basal diet with added 20% glucose, 20% MSG, 40% MSG, or 17% L-glutamic acid and sodium equivalent to 20% MSG for 5 weeks and then killed. No specific endocrine or neurological defects were seen (Wen et al., 1973).

Lesions of the arcuate nucleus and retina were produced by intraperitoneal administration of 4 g MSG/kg b.w. to rats on days 2, 4, 6, and 10 after birth followed by examination 3-5 months later. There was a reduction of cell bodies in the arcuate nucleus and degeneration of the visual system; reduced weights of the anterior pituitary and gonads were observed in adults (Clemens et al., 1978).

No CNS damage was observed in rats which were fed 4% MSG in the diet for 2 years or in dogs which were fed 10% MSG (Heywood & Worden, 1979).

Hamsters
Neonatal hamsters were given single subcutaneous injections of 8 g MSG/kg b.w. on days 6, 7, 8, 9, or 10 post partum and the brains were examined 6-12 hours later. Lesions of the arcuate nucleus were produced on days 6-8; dosing on days 9 or 10 still caused prominent lesions, but they were less severe than on previous days (Tafelski & Lamperti, 1977).

Groups of male and female hamsters aged 25 days were deprived of water or water and food overnight, then offered solutions containing 0, 2, 4, 6, or 8% MSG for 30 minutes. Six hours later the animals were killed and the brains examined; no hypothalamic lesions were seen (Takasaki & Torii, 1983).

Guinea pigs
A single subcutaneous injection of 1 g/kg MSG was given to 2- or 3-day old guinea pigs. Six animals were killed 3 hours after treatment and the hypothalamic area investigated by histochemical and histological techniques. No clear effect of MSG was discernible. With a dose of 4 g/kg, an increase in glial cells, vacuolization of cells in the arcuate nuclei, and some evidence of cell necrosis were observed.

The severity of the lesions was in no way comparable to the effects seen in the hypothalami of mice treated with the same doses (Golberg, 1973).

Chickens

Groups of 6 chickens were given subcutaneous doses of 4 g MSG/kg b.w. at days 5, 70, or 120 post-hatching. A further group of 120-day old birds was given subcutaneous doses of 7 g MSG/kg b.w. At 140 days there was no damage in the brains of birds injected at 120 days post-hatching, but lesions were seen in the hypothalami of the birds injected 5 days post-hatching and, less markedly, 70 days post-hatching (Robinzon et al., 1974).

Six groups of 14 birds were injected subcutaneously 5 days post-hatching with single doses of 1 or 4 g MSG/kg b.w. or with daily doses of 1 or 4 g MSG/kg b.w. on 10 consecutive days; controls received equivalent doses of saline. Forty days later, 3 birds from each group showed frequent necrotic neurons in the hypothalamus and a reduced-cell density (Snapir et al., 1971).

Ducks

Groups of 5-6 ducks aged 5 days were given either single subcutaneous injections of 1 or 4 g MSG/kg b.w. or 10 consecutive daily doses of 1 or 4 g MSG/kg b.w. and observed up to 235 days after injection. Brain lesions in the rotundus nuclei and the ventro-medial hypothalamic nuclei were noted in all birds; sperm mobility was low in birds with lesions in the mamillary bodies, but this probably was not due to MSG (Robinzon et al., 1975).

Rabbits

When rabbits were injected intraperitoneally with 0.25, 0.5, 1, or 2 g MSG/kg b.w. daily for 16 days, histology of the retinae showed degeneration of all layers. In the ERG the amplitudes of the a and b waves were decreased to less than half their normal values. The minimum effective dose for retinal degeneration was 0.25 g/kg b.w. (Hamatsu, 1964).

Dogs

Intravenous casein hydrolysate or synthetic amino acid mixture caused nausea and vomiting in dogs (Madden et al., 1944).

Groups of 3 dogs at 3 days or 35 days of age received subcutaneously or orally single acute doses of 1 g/kg b.w. monosodium glutamate, monopotassium glutamate, sodium chloride, sodium gluconate, or distilled water, and were sacrificed 3 or 24 hours after treatment. Preliminary light microscopic studies of the large midbrain area showed similar non-specific scattered tissue changes in all treated groups (Oser et al., 1971).

Groups of 6 pups, 3 to 4 days of age, were dosed either orally or subcutaneously with 1 g/kg b.w. monosodium glutamate, sodium chloride, sodium gluconate, potassium glutamate, or water. Pups were sacrificed at either 3 hours, 24 hours, or 52 weeks after dosing. Other groups of dogs 35 days of age received single doses of test material, and were sacrificed at either 4 or 24 hours post-dosing. Body weights of dogs which were dosed once at 3 days of age and followed for a year, showed no evidence of effects of any treatment. Femur weight, as well as weight of the pituitary gland, ovaries, uterus, and mammary glands were similar to controls. Gross and microscopic examinations of these tissues failed to reveal any abnormalities. Extensive microscopic examination of brain tissue of all test animals did not show any treatment-related changes (Oser et al., 1973).

Monkeys

Doses of 250 mg or 1 g/kg MSG were administered orally daily for 30 days to two groups of 3 infant rhesus monkeys starting at 1 day after birth. General clinical observations over a period of 30 days revealed normal growth, development, and activity. No changes were observed in the levels of haemoglobin, haematocrit, RBC or WBC counts, or reticulocytes. The levels of glucose, urea nitrogen, and serum potassium, calcium, and sodium were within the normal ranges. At autopsy, complete histological, histochemical, and ultrastructural investigations of the entire arcuate nuclei and median eminence region failed to reveal any necrotic or damaged neurons (Golberg, 1973).

A newborn (8 hours old) rhesus infant, probably somewhat premature, was given subcutaneously 2.7 g/kg b.w. monosodium glutamate. After 3 hours (no abnormal behaviour noted) the monkey was killed and the brain perfused in situ for 20 minutes. A lesion was seen in the periventricular arcuate region of the hypothalamus identical to those seen in mice given similar treatment. Electron microscopic pathological changes were seen in dendrites and neuron cells but not in glia or vascular components (Olney & Sharpe, 1969).

Monkeys, 4 days old, received single doses of monosodium glutamate (4 g/kg in phosphate buffer), either subcutaneously or orally. Animals receiving subcutaneous injections were sacrificed at 3, 24, and 72 hours, the one receiving an oral dose at 24 hours. No brain lesions were observed (Abraham et al., 1971).

Three infant monkeys, 5 days of age, received MSG by stomach tube at a dose of 2 g/kg. Two infant monkeys at 10 days of age, 2 at 20 days of age, and 2 at 40 days of age, received the same treatment. Two animals at 80 days of age received 4 g/kg. One control monkey was included in each group. The animals were observed for 4 hours after dosing and then sacrificed. After a period of fixation, a block of tissue was removed from each brain which included the hypothalamus. Serial sections, 10 mm thick, were made in the horizontal plane and examined by light microscopy. No changes that were considered to be associated with the administration of MSG were observed in the hypothalamus of the monkeys (Huntington Research Centre, 1971).

A group of 3 to 4 day-old cynomolgus monkeys received either subcutaneously or orally single doses of 1 g/kg b.w. monosodium glutamate, sodium chloride, sodium gluconate, or potassium glutamate and were sacrificed 3 or 24 hours post-dosing. Another group of monkeys (3 to 4 days old) received orally either 4 g/kg monosodium glutamate or sodium chloride, and were sacrificed at 3, 6, and 24 hours post-dosing (3 and 24 hours in the case of sodium chloride-dosed monkeys). Detailed microscopic examination of the hypothalamus did not show any evidence of monosodium glutamate-induced necrosis or any differences between any of the groups. Examination of the eyes did not reveal any effects due to monosodium glutamate. Glutamate and glutamine blood levels showed considerable variation in individual values among the animals dosed orally and subcutaneously. Subcutaneous dosing resulted in values an order of magnitude higher than those observed by oral dosing (Oser et al., 1973).

Monosodium glutamate was administered to 6 pregnant rhesus monkeys (Macaca mulatta) at a daily dosage equivalent to 4 g/kg b.w. during the last third of pregnancy. Four pregnant monkeys not receiving treatment were used as controls. Body weight and condition was unaffected throughout the gestation period. The duration of gestation was within the accepted range (156-178 days). There were no cases of delayed parturition or dystocia. Nursing, suckling, and behavioural patterns were normal except for one monkey which killed its infant at birth. Birth weights of the neonates were within the normal range. Infants, when removed from mothers, showed distress but no signs of abnormal behaviour. The hypothalamus region and related structure of the brain were examined by light microscopy. No abnormalities were observed (Heywood et al., 1972a).

Monosodium glutamate was administered by intragastric intubation at dosage levels of 2 g/kg to 2 monkeys aged 2 days. Two monkeys of similar age were used as controls. Four hours after dosing, the animals were sacrificed. Examination of the hypothalamus (bordered rostrally by the optic chiasma and caudally by the pons) by light and electron microscopy did not show changes caused by administration of the test compound. Changes observed by electron microscopy occurred as frequently in control animals as in test animals, and appeared to be due to fixation artifacts (Heywood et al., 1972b).

Rhesus monkeys (Macaca mulatta) aged between 4 and 80 days were divided into groups by age. Each group contained 3 test animals and 1 control animal. Dosage levels were 2 g/kg for animals up to 44 days of age and 4 g/kg for animals up to 80 days of age. The animals were observed for 4 hours and then sacrificed. Immediately prior to dosing and prior to sacrifice, serum and plasma samples were obtained for measurement of SGPT, SGOT, and plasma glutamic acid. At sacrifice liver samples were obtained for measurement of GPT and GOT. SGPT and SGOT values did not show significant increases over the test period. Plasma glutamic acid was within the normal range. Liver GPT and GOT values were within the normal ranges. Examination of the hypothalamus region and associated structures by light microscopy did not reveal any compound-related effects (Heywood et al., 1971).

Sixteen infant monkeys (M. mulatta or M. irus) were fasted for 4 hours before receiving by stomach tube single doses of a 50% solution of monosodium glutamate, equivalent to doses of 1, 2, or 4 g/kg b.w. Control animals received distilled water. At 6 hours post-dosing the animals were sacrificed and the brains perfused for examination by light and electron microscopy. No morphological differences were observed in the hypothalamic regions of treated and control monkeys. Inadequately fixed tissue had the same appearance as that of a previously-reported brain lesion in a newborn monkey (Reynolds et al., 1971).

Monosodium glutamate was administered into the circulation of primate fetuses (Macaca species) via the umbilical vein at dose levels of approximately 4 g/kg b.w. A total of 7 animals were treated. At time periods of 2 to 6 hours post-dosing the fetuses were delivered by caesarean section and the brains fixed for microscopic examination. The hypothalamic areas of the brain from all 7 fetuses were found to be completely normal. There was no evidence of pyknotic nuclei, tissue oedema, or neuronal loss in the arcuate region (Reynolds & Lemkey-Johnston, 1973).

Ten infant squirrel monkeys were fed either a 0, 4.8, 9.1, or 17% (based on dry weight) MSG formula diet for 9 weeks. Three of the test monkeys died. Two died of effects not related to MSG. The third, which was on the 17% diet, developed convulsive seizures. However, the other 2 animals in this group were unaffected. Clinical observations were made daily, and at the end of the test period the monkeys were sacrificed and the major organs examined microscopically. Sections of the retina and hypothalamus were examined by electron microscopy.   No hypothalamic or retinal lesions were observed (Wen et al., 1973).

In another study, an infant cynamologus monkey and an infant brush monkey were fed 0.1% MSG formula for one year. Daily observations revealed no behavioural abnormalities. Their weight gains, ERG, EEG, and plasma amino acids were similar to controls not consuming MSG. No evidence of gross obesity developed (Wen et al., 1973).

Groups of 3 infant monkeys were dosed with a mixture of water and skimmed milk containing either added NaCl or MSG, on an equivalent mole/kg basis. Administration was via nasogastric tube. Other groups were injected subcutaneously with either a 25% aqueous solution of MSG or a 10% solution of NaCl. The doses ranged from 1-4 g/kg b.w. All animals were sacrificed after dosing and the brains examined by combined light and electron microscopy. Infants given relatively low oral doses of MSG (1 and 2 g/kg) sustained small focal lesions confined primarily to the rostro-ventral aspect of the infundibular nucleus. Those treated with high subcutaneous doses developed lesions which spread throughout, and sometimes beyond, the infundibular nucleus. At all doses tested, and by either route of administration, rapid necrosis of neurons (within 5 hours) was observed. Measurements of blood glutamate levels suggested that the threshold for lesion formation in 1-week old rhesus monkeys may be in the range of 200 mg/l (Olney et al., 1972).

Neonatal primates given 1-4 g MSG/kg b.w. orally showed only elevated aspartate and glutamate blood levels; no hypothalamic lesions were noted (Boaz et al., 1974).

Three monkeys were given orally 2 g MSG/kg b.w. at 3, 60, or 99 days of age; a fourth monkey aged 180 days served as a control. After 4 hours, examination of the brains revealed no abnormalities. A further 16 monkeys were divided into five groups; four groups received orally 2 g MSG/kg b.w. and one group was given 4 g/kg b.w. Examination of the brains and of serum GOT and GPT showed no abnormalities (Newman et al., 1973).

Six pregnant monkeys were given 4 g MSG/kg b.w. for the last third of pregnancy and a further 4 untreated animals served as the source of control offspring. The offspring were killed 4 hours after birth, when no abnormalities in histopathology of the brains were seen. In another experiment, two monkeys aged 2 days were given 2 g MSG/kg b.w. and the brains were examined after 4 hours. No abnormalities were seen (Newman et al., 1973).

No treatment-related hypothalamic lesions were observed in neonatal rhesus monkeys killed 3 or 24 hours or 8, 15, or 30 days following oral or subcutaneous administration of 0.25, 1, or 4 g MSG/kg b.w. The hypothalamus was examined by light and electron microscopy and the arcuate nuclei, median eminence, ependymal and glial cells were comparable to controls (Abraham et al., 1975).

Neonatal monkeys aged 1-14 days were given 1-4 g MSG/kg b.w. by gavage. No lesions were detected in the hypothalamus (Reynolds et al., 1976). No evidence of hypothalamic lesions was seen following in vitro exposure in 1 embryonic and 7 fetal brains (Reynolds et al., 1979).

Ten neonatal squirrel monkeys were divided into four groups of 2-3 animals; 2 received infant formula, 3 received infant formula with 5% added MSG, 2 were given infant formula with 10% added MSG, and 3 were given infant formula with 20% added MSG from day 11 to day 21 post partum followed by observation for 9-10 weeks. The animals were killed in the twelfth week. No abnormalities were observed in EEG or ERG scans and no retinal or hypothalamic lesions were detected (Wen et al., 1973).

One cynomolgus monkey and 1 bushbaby were given 10% MSG in their diets from the first to the eleventh month of age. No adverse effects were noted on growth or on EEG or ERG scans. In another experiment, a 3-week old cynomolgus monkey was injected intramuscularly with 2.7 g MSG/kg b.w. and observed for 2.5 years. No untoward effects were seen (Wen et al., 1973).

The administration of MSG to a 2-day old rhesus monkey at a dose level of 4 g/kg b.w. in baby formula failed to induce any pathological changes in the hypothalamus (Heywood & James, 1979).

Special studies on postnatal behaviour

Mice

Neonatal mice aged 2-11 days were injected with increasing doses of 2.2 to 4.4 g MSG/kg b.w. over 10 days. Examination at 2, 10, and 20 weeks after weaning showed no increase in food consumption, but obesity was observed. Locomotor activity was depressed at all three time-points, and oxygen consumption was decreased at 10 and 20 weeks. There were no effects on plasma thyroxine levels (Poon & Cameron, 1978).

Rats

Eight litter mates from each of 10 pregnant Holtzmann rats were divided into four groups consisting of two litters from each mother. Litters of these groups received water (control) and MSG solution (1.25, 2.5, or 5 g/kg) by stomach tube daily from days 5 to 10 of age. At day 21 the rats were placed in separate cages and at 3 months of age they were subjected to three different behavioural situations, namely, spontaneous motor activity, T-maze, and fixed-ratio food reinforcement. Rats in the high-dose group showed less spontaneous motor activity than the controls and a deficiency in discrimination learning in the T-maze study. However, learning of a fixed-ratio food reinforcement schedule was not affected. After 3 months weight gain of the treated animals was less than controls, the effect being greatest in the 5 g/kg dose group (Pradhan & Lynch, 1972).

Rats given 4 g MSG/kg b.w. on days 1-10 of life and tested at 50 days in a swimming maze were less able to learn the maze. Glutamate levels in the brain, liver, and blood were raised after 10 days, while other amino acid levels were also changed. Structural alterations were probably responsible for the permanent impairment of brain function (Berry et al., 1974).

Rats given 10 g MSG/kg b.w. orally showed depressed avoidance acquisition and shuttlebox test performance. True pharmacological tolerance rapidly appeared (Pinto-Scognamilio et al., 1972).

A group of 16 neonatal rats was given MSG between day 3 and day 8 in progressively increasing subcutaneous doses of 2.5 to 4.2 g/kg b.w.; a further 16 animals acted as controls. At weaning, half of each group was given exercise wheels. The animals were killed at 110-112 days post-weaning. No effects of MSG were observed on motor activity but both sexes were obese despite exercise and hypophagia. Disturbances of the female reproductive system and arrested skeletal development were noted (Nikoletseas, 1977).

Rats (174 males and 196 females) were divided according to treatment as follows: three groups 2 days of age were injected subcutaneously with saline, 0.2 g MSG/kg b.w., or 4 g MSG/kg b.w. for 10 days; one group 10 days of age was injected subcutaneously with 4 g MSG/kg b.w. for 10 days; two groups 10 days of age were given saline or 0.5 g MSG/kg b.w. by gavage; and one group 10 days of age was put on an ad libitum diet containing 10% MSG. The animals were observed for up to 9 months. Multiple injections of 4 g MSG/kg b.w. to neonates caused low grip strength, hypoactivity, changes in spontaneous motor activity, deficit of learning ability, and tail mutilation. The same treatment beginning at 10 days of age resulted in only slight behavioural abnormalities later in life or no detectable changes. Administration of subneurotoxic doses either by subcutaneous injection or gavage, or in high levels in the diet, were without behavioural effects. Adverse effects were not observed when the brains were free from histological evidence of injury (Iwata et al., 1979).

Neonatal male rats were given 4 consecutive daily subcutaneous injections of 4 g MSG/kg b.w. on days 1 to 4 post partum. When adult studies of sleep patterns indicated an increase in total sleep duration with more pronounced effects on paradoxical sleep due to treatment, circadian rhythmicity tended to degenerate into ultradian (6, 8, 12 hours) harmonics. There was an almost complete disappearance of ACTH and α-MSH immunoreactive perikarya in the rostal part of the arcuate nucleus (Olivo et al., 1986).

Chickens

Applying 7.5 mg MSG/kg b.w. to forebrains of chicks once in the first week after hatching caused learning defects (Rogers, 1982). However, intercranial applications are irrelevant to the evaluation of orally-ingested MSG and learning in birds depends on motivation, which was not described in this experiment (Wurtman, 1983).

Special studies on reproduction and teratogenicity
Mice

Groups of 6 4CS or Swiss white mice (3 males and 3 females) were maintained on diets containing 0, 2 (equal to 4 g/kg b.w./day), or 4% (equal to 8 g/kg b.w./day) monosodium glutamate. Mice were mated after 2 to 4 weeks on the test diets. F_1 offspring were weaned at age 25 days and fed the same diet as the parents. At age 90 days, selected F_1 male and female mice from each group were allowed to produce a single F_2 litter.

Parent mice were maintained on test diets for 100 days after delivery and F_1 mice for 130 days after delivery. F_2 mice were reared until 20 days of age. No effects were observed on growth, feed intake, oestrous cycle, date of sexual maturation (F_1 generation), organ weights, litter sizes, body weights of offspring, or histopathology of major organs (including brain and eyes) of the parent and F_1 generations. Day of eye opening, general appearance, and roentgenographic skeletal examinations of F_2 generation animals showed no abnormalities (Yonetani et al., 1970).

Five groups of 24-30 mice received 0, 5.2, 24, 112, or 520 mg MSG/kg b.w. for 10 days during pregnancy. No clear adverse effects were seen on nidation or on maternal or fetal survival. There were no adverse effects on resorptions, fetal weights, or litter parameters, and no differences were noted in soft tissue or skeletal abnormalities (Food and Drug Research Laboratories, 1974a).

Neonatal mice were injected with increasing daily doses of MSG from 2.2 to 4.2 g/kg b.w. over 10 days and observed subsequently for up to 302 days. Treated females had fewer pregnancies and smaller litters than the controls; males displayed reduced fertility. The body weights of both sexes were increased while organ weights of the pituitary, thyroid, and ovaries/testes were reduced (Pizzi et al., 1977).

Six groups of mice were given 0, 1, or 2% MSG in a diet supplemented with 1 or 2% vitamin mix. Females from the F_1 (33) and F_2 (29) generations were observed for reproductive function. Animals receiving MSG showed higher weaning weights, better survival rates, and no effects on brain cellularity (Semprini et al., 1974a).

A multi-generation reproduction study on MSG was conducted on IVCS and Swiss albino mice. Groups of 3-5 60-day old male and female mice were maintained on diets containing 0, 2, or 4% MSG from 2 weeks prior to mating until 100 days after parturition. Animals in the F_1 generation were maintained on the same diets and mated at 90 days of age. Animals in the F_2 generation were killed on day 20. No significant abnormalities were observed on growth, food consumption, oestrus cycle, date of sexual maturation, organ weights, litter sizes, body weights of offspring, or histopathology of major organs (including brain and retina) of the parents and the F_1 generation. Mice of the F_2 generation showed normal date of eye-opening. No teratogenic effects were observed (Yonetani et al., 1979).

In a 3-generation reproduction study, two groups of 17 male and 51 female CD-1 COBS mice were given 1 or 4% MSG in the diet; a further control group of 33 male and 99 female animals received diets without MSG. Animals in the F_1 and F_2 generations were sacrificed at 27-36 weeks of age, while some of the F_3 generation animals were examined histopathologically at 0, 3, 14, and 21 days. Growth and food intake were similar in all groups. The actual MSG intakes were 1.5 and 6 g/kg b.w./day for males and 1.8 and 7.2 g/kg b.w./day for females in the 1 and 4% treatment groups respectively. The MSG intake of dams rose markedly during lactation, rising to a maximum of 25 g/kg b.w./day.

No adverse effects were noted on fertility, gestation, viability, or lactation indices of progeny of any generation, and no brain lesions or other treatment-related histopathology were observed (Anantharaman, 1979).

Rats

Six groups of 5-6 male and 5-10 female rats received by oral intubation daily 25 or 125 mg/kg b.w. glutamic acid monohydrochloride. Males and females received the compound during days 5-19 of the first month, days 20-31 of the following month, and days 1-10 during the third month. No adverse effects were noted on weight gain, feed intake, or sexual cycles of females. No organ-weight changes were seen in females but males at the higher-dose level had enlarged spleens. Animals were mated at the end of the experiment and the pups were normal (Furuya, 1967).

Rats were given thalidomide combined with 2% L-glutamic acid, and showed essentially the same defects in the pups as groups treated with thalidomide alone. A group receiving L-glutamic acid alone was not different from controls (McColl et al., 1965).

Four females and 1 male fed for 7 months 0, 0.1, or 0.4% monosodium L-glutamate, monosodium DL-glutamate, or L-glutamic acid were mated. The number of pups per litter was similar in all groups. Only 15-20% survived because of cannibalism. No abnormalities regarding fertility were seen after mating other groups of 4 females and 1 male at 9 and 11 months. The F_1 generation was mated at 10 months and an F_2 generation was produced in most groups, but only the groups fed 0.1 and 0.4% L-glutamic acid produced F_3 and F_4 generations. No impairment of fertility was noted (Little, 1953a).

Monosodium glutamate was administered orally at doses up to 7 g/kg/day to pregnant rats on days 6-15 or 15-17 following conception. The substance produced no adverse effect in the progeny up to the period of weaning. Further physical development to maturity was also normal except that the progeny obtained from gravida treated on days

15-17 during gestation showed impaired ability to reproduce (Khera et al., 1970).

Two female rats received 4 g/kg b.w. monosodium glutamate commencing at day 1 of pregnancy. There were no effects on pregnancy or lactation. Pups were divided into three groups. Two groups were nursed by parents receiving monosodium glutamate, and one group by untreated parents. At weaning (day 20), one group of pups that had been nursed by a parent receiving monosodium glutamate received approximately 5 g/kg monosodium glutamate daily for 220 days. Parents received 4 g/kg monosodium glutamate for 336 days. No effects were observed on growth or the oestrus cycle. All pups developed normally, and no abnormalities were noted in growth rate, time of sexual maturity, oestrus cycle, or fertility. For histological studies, the brain, hypophysis, and eye were fixed in 10% neutral buffered formalin. Sections were stained with haematoxylin-eosin and Luxol fast blue-cresyl violet. No differences were observed in the arcuate nuclei, medium eminence of the hypothalamus, or retina between control and monosodium glutamate-treated groups (Suzuki & Tagahashi, 1970; Shimizu & Aibara, 1970).

Groups of female rats were maintained on diets containing 0.5, 1, or 2% vitamin mix. At each vitamin level diets also contained monosodium glutamate at 0, 1, or 2%. Reproductive performance of the parental rats as well as of the F_1 offspring maintained on similar diets was studied. The addition of monosodium glutamate to the diet resulted in an increased fertility rate as well as increased survival at weaning of the offspring of the F_1 generation. Addition of monosodium glutamate to the diet had no effect on growth rate in the neonatal period. Analysis of the brain tissue of first and second generation offspring at birth for RNA, DNA, protein, nucleus number, and cellular size showed that the brains of rats born of parental mothers on monosodium glutamate diets contained a smaller number of nuclei and larger cells than controls. In contrast, offspring of the F_1 generation showed increased RNA, DNA, and nucleus numbers when compared with the offspring of the parental generation. The differences present at birth disappeared at weaning (Semprini et al., 1971).

Parental rats were fed 0 or 10% MSG in the diet, mated, and F_1 pups were maintained for 100 days on the same diet as the parents, then mated. Of the F_2 pups, 10 were sacrificed on each of days 1, 2, 3, 5, 10, and 21 post partum. No effects on reproductive function were observed as indicated by conception rates and numbers of pups per litter, and no differences were noted in post-natal development, as measured by brain, liver, and body weights. Brain and liver glutamate, aspartate, protein, DNA, RNA, and the activity of glutamic acid decarboxylase were unaffected by treatment (Prosky & O'Dell, 1972).

Groups of 25 pregnant Wistar-derived rats were given 0, 4.5, 21, 97, or 450 mg monopotassium glutamate/kg b.w. by oral intubation on days 6-15 of pregnancy. No effects were observed on nidation or on maternal of fetal survival and the number of abnormalities in the offspring in the test groups did not differ from those occurring spontaneously in controls (Food and Drug Research Laboratories, 1974b).

Rats were given daily s.c. doses of 2 or 4 g MSG/kg b.w. from days 2-10 post partum or a dose of 4 g/kg b.w. for 10 days from day 10. Other groups of 10-day old rats received 0.5 g MSG/kg b.w. orally for 10 days or similar doses followed by weaning on to a diet containing 5% MSG. Females repeatedly treated as neonates showed precocious puberty, disturbed oestrus cycles, small ovaries and pituitaries, and a poor response to gonadotrophin. Females treated from day 10 developed normal puberty and oestrus cycles, but later cycling became irregular. Low doses of MSG given orally or by repeated injection had no effect (Matsuzawa et al., 1979).

Hamsters

Groups of neonatal hamsters were given subcutaneous injections of 0, 4, or 8 g MSG/kg b.w. in saline on days 1-5, 6-10, or 1-10 postnatally. The animals were sacrificed on day 60 (males and acyclic females) or on a day of ovulation near day 60 (regular cycling females). MSG-treated animals had significantly lower reproductive organ weights than controls. Lesions were detected in the arcuate nucleus only in hamsters receiving 8 g MSG/kg b.w. on days 6-10 or 1-10; female hamsters were acyclic, and had ovaries with small follicles and no corpora lutea. Administration of 50 i.u. pregnant mare's serum to these animals caused follicular maturation, and ovulation occurred after administration of 10 i.u. human chorionic gonadotrophin (HCG). Males had atrophic seminiferous tubules which recovered normal histology and steroid dehydrogenase activity after treatment with HCG. The evidence indicated that MSG had affected the hypothalamic centres controlling GSH and LH release (Lamperti & Blaha, 1976).

Male and female hamsters were given saline or 8 g MSG/kg b.w. by subcutaneous injection on days 7 and 8 postnatally. As adults, all MSG-treated females were acyclic, had significantly lower uterine and pituitary weights, and lower levels of FSH in the plasma and anterior pituitary from controls. The ovaries had small follicles and the interstitial cells were hypertrophied. It was reported that only 7.3% of the neurons of the arcuate nucleus were morphologically intact. Adult male hamsters that had been given MSG neonatally had a treatment-related reduction in testicular, seminal vesicle, and pituitary weights as well as lower FSH levels in the plasma relative to controls. Approximately 14% of the neurons in the arcuate nucleus were morphologically intact. The seminiferous tubules were histologically atrophic in only 3 out of 8 animals. These results were taken as supporting previous reports that MSG lesions of the arcuate nucleus result in alterations in FSH, but not LH, secretion in the hamster (Lamperti & Blaha, 1980).

Chick embryos
Fertilized hen eggs were incubated after single injections of 0.01-0.1 mg glutamic acid into the yolk sac. The mortality of embryos was raised compared with controls (53% against 24%) and there was a higher incidence of developmental defects (24% against 3%), especially depression of development of the spine, pelvis, and lower limbs (Aleksandrov et al., 1965). In another study, many variables were studied such as the route of injection, dose, and time of injection. No obvious toxic or teratogenic effects were observed (US FDA, 1969).

Rabbits

In one group of 10 female and 4 male rabbits, only the females received 25 mg/kg b.w. glutamic acid for 27 days. Two of the females were pregnant and the others were not pregnant. A second group of 4 females and 2 males received 25 mg/kg glutamic acid along with 25 mg/kg vitamin B_6. A third group of 6 females and 2 males received 25 mg/kg glutamic acid alone. A fourth group of 20 females and 8 males served as controls. The test substance was given by gavage. The first group showed 2 animals with delayed pregnancy, the uterus containing degenerated fetuses. Two others had abortions of malformed fetuses. Two animals delivered at the normal time, but the pups had various limb malformations. Four animals did not conceive. The pups did not become pregnant before 7 months of age and showed limb deformities, decreased growth, and slow development compared with controls. Histopathological examination showed scattered atrophy or hypertrophy of different organs. The second group produced 2 pregnant females which delivered malformed pups. These died soon after birth and showed bony deformities as well as atrophic changes in various organs. The third group produced 3 pregnant females which delivered pups with limb deformities. All three groups showed testicular atrophy in parents and multiple changes in the pups (Tugrul, 1965).

Four groups of rabbits (24 females and 16 males) received 0, 0.1, 0.825, or 8.25% monosodium glutamate in their diet for 2 or 3 weeks before mating. A positive control group of 22 pregnant females received 100 mg/kg thalidomide from days 8 to 16 of pregnancy. All does were sacrificed on days 29 or 30 of gestation and the uteri and uterine contents were examined. All males were sacrificed and the gonads and any abnormal organs examined. No significant effects on body-weight gain, food consumption, general appearance, or behaviour were observed. Gross and histopathological examinations revealed no toxic effects on embryos or resorptions. Pups and all litter data were comparable among test animals and negative controls. The brains of 5 female and 5 male pups at the 8.25% level were subsequently checked for neuronal necrosis compared with controls, but none was found. Similar investigations on 5 male and 5 female pups at the 0.1 and 0.825% levels were also negative (Hazelton Laboratories, 1966, 1969a,b).

In another experiment in rabbits, animals received 2.5, 25, or 250 mg/kg b.w. L-glutamic acid hydrochloride at 70 and 192 hours post coitum. Operative removal of fetuses was performed on the eleventh, seventeenth, and thirtieth day post coitum in three different series. The corpora lutea and the resorbed and implanted normal and deformed fetuses were examined. No significant effects due to L-glutamic acid were noted with respect to teratogenesis (Gottschewski, 1967).

Glutamic acid hydrochloride at a dose of 25 mg/kg b.w. was given orally to 15 pregnant rabbits once a day for a period of 15 days after copulation; monosodium glutamate at the same dose and for the same period of time was given to 9 pregnant rabbits, and saline solution was administered to 11 pregnant rabbits which served as a control group. No differences were noted between the treated groups and the controls with regard to the rate of conception, mean litter size, or nursing rate. The average body weights of the young in the treated groups were slightly lower as compared with the control group, but the weights of testes, ovaries, and adrenal glands in the young and ovaries, adrenal glands, liver, kidneys, and spleen in the mothers were not different from those in the controls. In the young, no external or skeletal malformations were observed. There were some abnormal changes in gestation such as abortion or resoption of fetuses, but these observations were made in all groups, with incidences of 21% in the glutamic acid hydrochloride group, 25% after administration of monosodium glutamate, and 27% in the controls. There were no external or skeletal malformations in the aborted fetuses (Yonetani, 1967).

Special studies on pharmacological effects

Intravenous injection of large doses of glutamic acid in rabbits caused ECG changes that could be interpreted as symptoms of myocardial lesions. Arterial hypertension induced by glutamic acid preparations was demonstrated to be of central origin. Studies with isolated heart showed that large doses of glutamic acid slowed heart action, increased systolic amplitude, and constricted coronary vessels. Very large doses stopped cardiac action (Mazurowa et al., 1962).

Ten male and 4 female subjects were given orally either 25 or 250 mg MSG/kg b.w. alone, 250 mg MSG/kg b.w. together with atropine, or prostigmine, and plasma cholinesterase activity was measured. There was little effect with the lower dose or the higher dose plus atropine. The high-dose effect of MSG alone was similar to the effect of prostigmine. Hence, large doses may cause release of an acetylcholine-like substance acting on the parasympathetic system in so-called "Chinese restaurant syndrome" (Ghadimi et al., 1971).

Acute toxicity

The acute toxicity of glutamate by various routes of administration in several animal species is given in Table 3.

Short-term studies
Mice

Thirty-eight neonatal mice were observed for 9 months. Twenty received monosodium glutamate by subcutaneous injection daily for 1 to 10 days after birth at doses of 2.2-4.2 g/kg. Eighteen were used as controls. Even though the treated animals remained skeletally stunted and both males and females gained more weight than controls from 30 to 150 days, the treated animals consumed less food than controls. The test animals were lethargic and the females failed to conceive, but male fertility was not affected. At autopsy, massive fat accumulation, fatty livers, and thin uteri were observed in test animals, and the adenohypophysis had fewer cells overall (Olney, 1969b).

Ten test neonates received single subcutaneous injections of 3 g/kg monosodium glutamate 2 days after birth. Thirteen neonates were used as controls. Test animals were heavier than controls after 9 months, but less so than mice given repeated injections in the experiment described above. The author postulated that an endocrine disturbance could lead to skeletal stunting, adiposity, and female sterility. Lesions differed from those due to gold thioglucose or bipiperidyl mustard, which affect the ventro-medial nucleus and cause hyperphagia (Olney, 1969b).

Table 3. Results of acute toxicity assays on glutamate

Species	Route	LD_{50} (mg/kg b.w.)	Reference
Mouse	oral	12,961	Izeki, 1964
	oral	16,200 (14,200-18,400)	Ichimura & Kirimura, 1968
	oral	19,200 (16,130-22,840)	Pinto-Scognamiglio et al., 1972
	s.c.	8,200 [1]	Moriyuki & Ichimura, 1978
	i.p.	6,900	Yanagisawa et al., 1961
	i.v.	30,000	Ajinomoto Co., 1970
Rat	oral	19,900 (L-MSG)	International Min. & Chem. Corp., 1969
	oral	10,000 (DL-MSG)	International Min. & Chem. Corp., 1969
	oral	16,600 (14,500-18,900)	Pinto-Scognamiglio et al., 1972
	s.c.	8,200 [1]	Moriyuki & Ichimura, 1978
Guinea-pig	i.p.	15,000	Ajinomoto Co., 1970
Rabbit	oral	> 2,300 (L-GA)	International Min. & Chem. Corp., 1969
Cat	s.c.	8,000	Ajinomoto Co., 1970

[1]   No sex differences were noticeable in the toxic signs; high doses led to excitation.

Rats

Natural monosodium L-glutamate, synthetic monosodium L-glutamate, and synthetic monosodium D-glutamate in amounts of 20, 200, or 2000 mg/kg b.w. were given orally to groups of 5 male rats once a day for a period of 90 days. No effects on body weight, growth, volume and weight of cerebrum, cerebellum, heart, stomach, liver, spleen, or kidneys in comparison with the control group were observed. No histological changes in internal organs were found by macroscopic and microscopic examinations (Hara et al., 1962).

Male Sprague-Dawley albino rats were allowed water ad libitum and fed ground laboratory chow having 24% protein content. High levels of single amino acids in the diet of rats beginning at 21 days produced decreased food intake and a severe growth depression. These effects were dependent upon the kind and concentration of supplemented amino acid. L-Methionine caused the most severe growth depression while L-phenylalanine, L-tryptophan, and L-cysteine were also severely toxic. Less toxic were L-histidine, L-lysine, and L-tyrosine.

All other amino acids tested, including glutamic acid, had only a slight effect on growth, or none at all. Growth depression was attributed both to depressed food intake and to specific toxic effects of the amino acids. A direct correlation was not found between the toxicity of any dietary amino acid and its concentration in the blood (Daniel & Waisman, 1968).

Nine groups of 20 rats were given 0.5 or 6% calcium glutamate in their diet. No effect was noted on maze learning or recovery from ECT shock (Porter & Griffin, 1950).

Two groups of 14 rats received 200 mg L-monosodium glutamate per animal for 35 days. No differences in their learning ability for maze trials were noted (Stellar & McElroy, 1948).

Eight male rats fed 5% dietary DL-glutamic acid in a low protein diet (6% protein) showed little or no depression of growth, when compared to low protein controls. There was a 50% increase in the free glutamic acid in the plasma (Sauberlich, 1961).

Long-term studies
Mice

One control group of 200 male mice and six test groups of 100 male mice received 1 or 4% L-glutamic acid, monosodium L-glutamate, or DL-monosodium glutamate in their diet. No malignant tumours appeared after 2 years that could be related to the administration of test material. Growth and haematology were normal, and histopathological examination showed no abnormalities in the test animals (Little, 1953a).

Six groups of C57Bl mice, comprising 50 males and 50 females, were given diets containing 1 or 4% L-glutamic acid, L-MSG, or DL-MSG for 715 days. A further control group of 100 animals of each sex received basal diet. No treatment-related differences were seen in mortality, body-weight gain, incidence of concurrent disease, haematology, or tumour incidence (Ebert, 1979a).

Rats

Groups of 75 male and 75 female rats received for 2 years dietary levels of 0, 0.1, or 0.4% monosodium L-glutamate, monosodium DL-glutamate, or L-glutamic acid. No adverse effects were noted on body weight, growth, food intake, haematology, general behaviour, survival rate, gross or histopathology, or tumour incidence (Little, 1953b).

Six groups of Sprague-Dawley rats, comprising 35 animals of each sex, were fed diets containing 0.4 or 4% L-glutamic acid, L-MSG, or DL-MSG from 12 weeks to 2 years of age; a control group of 61 males and 69 females received basal diets. The protocol included a reproduction phase. No adverse effects were noted on behaviour, body-weight gain, food consumption, motor activity, clinical observations, haematology, or tumour incidence. Fertility, survival of the young, organ weights, and histopathology were comparable between controls and test animals (Ebert, 1979b).

Groups of 40 rats of each sex were given diets containing 0, 1, 2, or 4% MSG or 2.5% sodium propionate (positive control) for 104 weeks. Animals of each sex per group were examined at 12 weeks with full histopathology. Ophthalmological examinations every 13 weeks were negative and no adverse effects were noted on body weight, food consumption, haematology, blood chemistry, terminal organ weights, or survival rates. Tissues from 25 organs were examined histologically. Water consumption, urinary volume, and sodium excretion were increased at the 4% MSG level and sub-epithelial basophil deposits were observed in the renal pelvis. Focal mineral deposits from the renal cortico-medullary junction were equally distributed in all groups (Owen et al., 1978a).

Dogs

Beagle dogs were fed diets containing 0, 2.5, 5, or 10% MSG or 5.13% sodium propionate (control) for 104 weeks. There were no adverse effects on body-weight gain, food consumption, behaviour, ECG, ophthalmology, haematology, blood chemistry, organ weights, or mortality due to treatment. Urinary volume and sodium excretion were slightly elevated in animals receiving MSG or sodium propionate but kidney function was unimpaired. No treatment-related histological changes were observed (Owen et al., 1978b).

Observations in man

Intravenous glutamic acid (100 mg/kg b.w.) produced vomiting (Madden et al., 1944). The occurrence of nausea and vomiting following the intravenous administration of various preparations in a series of 57 human subjects was found to parallel the free glutamic acid content of the mixture. There was a direct relationship between free serum glutamic acid and the occurrence of toxic effects following intravenous administration. When serum glutamic acid reached 12-15 mg/100 ml, nausea and vomiting occurred in half the subjects. Other amino acids appeared to potentiate the effect (Levey et al., 1949).

Arginine glutamate may be used in the treatment of ammonia intoxication. It is given by intravenous infusion at doses of 25 to 50 g every 8 hours for 3 to 5 days in dextrose and infused at a rate of no more than 25 g over 1 or 2 hours; a more rapid infusion may cause vomiting (Martindale, 1967).

Monosodium glutamate has been used in the treatment of mentally-retarded children at doses up to 48 g daily, but on average at doses of 10-15 g. One hundred and fifty children aged 4-15 years were treated with glutamic acid for 6 months and compared with 50 controls. No significant rise in intelligence quotient was observed, but 64% were claimed to show improved behavioural traits (Zimmerman & Burgemeister, 1959).

Seventeen patients received up to 15 g monosodium glutamate 3 times a day, but they showed a raised blood level for 12 hours only.

No effects were noted on BMR, EEC, ECG, blood pressure, heart rate, respiration rate, temperature, or weight over a period of 11 months (Himwich, 1954).

Fifteen grams, then 30 g, monosodium glutamate were given for one week each, followed by 45 g for 12 weeks, to 53 patients. There were no effects on basal plasma levels of glutamic acid (Himwich et al., 1954).

DL-Glutamic acid•HCl was given at doses of 12, 16, or 20 g to eight patients with petit mal and psychomotor epilepsy without adverse effects (Price et al., 1943). Five episodes of hepatic coma in 3 patients were treated with 23 g monosodium glutamate i.v. with improvement (Walshe, 1953). L-Glutamic acid, 10 to 12 g given to epileptics and mental defectives, appeared to improve 9 out of 20 cases (Waelsch, 1949).

In studies of long-term health effects of MSG, no increased neurological symptoms and less myocardial infarction and stroke were seen. Blood glucose, cholesterol, and obesity were unrelated to MSG intake (Go et al., 1973).

Six women with well-established lactation patterns were fasted overnight and given single oral doses of 6 g MSG in water or in liquid diet; 4 controls received lactose. Milk samples were obtained 1, 2, 3, 4, 6, and 12 hours after administration; blood samples were collected 0, 30, 60, 120, and 180 minutes after administration of MSG or lactose. Small increases in plasma glutamate, aspartate, and alanine levels were noted, but little change was observed in breast milk amino acid levels (Stegink et al., 1972).

Following early anecdotal reports of subjective symptoms after ingestion of Chinese meals (Kwok, 1968) and experiments involving administration of MSG (Schaumberg et al., 1969; Kandall, 1968; Beron, 1968), MSG was implicated as the causative agent of "Chinese restaurant syndrome". Extensive studies in human volunteers have been carried out subsequently and have been reviewed recently (Kenney, 1986). These studies have failed to demonstrate that MSG is the causal agent in provoking the full range of symptoms of Chinese restaurant syndrome. Properly-conducted double-blind studies among individuals who claimed to suffer from the syndrome did not confirm MSG as the causal agent. Food symptom surveys have been considered technically flawed because of questionnaire design (Kerr et al., 1977, 1979a, 1979b).

Twenty-four subjects, including 18 who had a history of subjective flushing symptoms after eating Chinese restaurant food, were challenged with 3-18.5 g MSG. No one reported flushing sensations. Six subjects, 3 with a history of flushing, were challenged with 35-285 mg MSG/kg b.w. or 7.1-7.4 mg pyroglutamate/kg b.w. None reported flushing sensations and significant changes in facial cutaneous blood flow were not recorded (Wilkin, 1986).

COMMENTS

Food additive uses of MSG include its incorporation into food and its use as a condiment. Glutamic acid is a component of proteins and comprises some 20% of ingested protein. Bound glutamate is released during digestion in the lumen of the gut and in the mucosa. Transamination to alanine occurs during gastrointestinal absorption with concomitant formation of α-ketoglutarate; glutamine and glutathione are other metabolic products. After deamination, excess glutamic acid may be utilised in gluconeogenesis. As a consequence of transamination, there is only a small rise in portal glutamate levels unless very large doses are administered. Further metabolism occurs in the liver and only when this system is overwhelmed does the level of glutamate in systemic circulation rise significantly. Human infants, including premature infants metabolize glutamate similarly to adults.

High oral doses of MSG by gavage (in excess of about 30 mg/kg b.w.) or parenteral administration of MSG may lead to elevated blood levels. Administration with food, particularly metabolizable carbohydrate, lowers the peak plasma levels attained; peak plasma levels are also concentration-dependent and limited by unpalatability at high concentrations.

Oral ingestion of large amounts of glutamate does not increase concentrations of glutamic acid in maternal milk and, at least in rats and monkeys, glutamate does not readily cross the placental barrier.

Reproducible lesions of the CNS have been produced in rodents, lagomorphs, and primates following parenteral administration of glutamate or after forced intubation of very high doses. CNS lesions have never been observed after ad libitum administration of high concentrations of MSG in the diet or drinking water, except following dehydration by water deprivation in mice.

There are species, strain, and age-dependent differences in sensitivity to neuronal injury, neonatal mice being most sensitive and rats, guinea pigs, and primates less so. The threshold plasma levels for neuronal damage in the mouse, the most sensitive species, are 100-130, 380, and > 630 μmoles/dl in infant, weanling, and adult animals, respectively. In human studies, plasma levels of this magnitude have not been recorded even after ingestion of a single dose of 150 mg MSG/kg b.w. in water.

The oral ED_{50} for production of hypothalamic lesions in neonatal mice is approximately 500 mg MSG/kg b.w., while the largest palatable dose is about 60 mg MSG/kg b.w.

Acute, short-term, and chronic toxicity studies on MSG in the diet of several species have not shown specific toxic effects and there is no evidence of carcinogenicity or mutagenicity. Reproduction and teratologic studies using the oral route have been uneventful, even when the parental generation is fed glutamate at high doses, suggesting that the fetus or suckling neonate is not exposed to toxic levels through the maternal diet.

Controlled, double-blind crossover trials have failed to demonstrate an unequivocal relationship between "Chinese restaurant syndrome" and consumption of MSG. MSG has not been shown to provoke bronchoconstriction in asthmatics.

Caution should be used when ingesting MSG as a large single dose rather than divided between several meals because high plasma levels may be reached under the former conditions.

In its previous evaluation, the Committee concluded that it would be prudent not to apply the ADI for glutamate to infants under

12 weeks of age (Annex 1, reference 32). In view of the finding that infants metabolize MSG in a similar way to adults, no additional hazard to infants was indicated. However, the present Committee expressed the general opinion that the use of any food additives in infants foods should be approached with caution.

EVALUATION
Estimate of acceptable daily intake for man
ADI "not specified".

1. This is a group ADI for L-glutamic acid and its ammonium, calcium, magnesium, monosodium, and potassium salts.

2. Substances given an ADI "not specified", such as glutamate salts in this instance, are of low toxicity. On the basis of the available data (chemical, biochemical, toxicological, and other), the total dietary intake of glutamates arising from their use at the levels necessary to achieve the desired technological effect and from their acceptable background in food do not, in the opinion of the Committee, represent a hazard to health. For that reason, the establishment of an acceptable daily intake expressed in numerical form is not deemed necessary. The Committee reiterated the general principle expressed in its first report (Annex 1, reference 1) that the amount of an authorized additive used in food should be the minimum necessary to produce the desired effect.

REFERENCES

Abraham, R., Dougherty, W., Golberg, L., & Coulston, F. (1971). The response of the hypothalamus to high doses of monosodium glutamate in mice and monkeys. Exp. Mol. Pathol., 15, 43-60.

Abraham, R., Swart, J., Golberg, L., & Coulston, F. (1975). Electron microscopic observations of hypothalami in neonatal rhesus monkeys (Macaca mulatta) after administration of monosodium L-glutamate Exp. Mol. Pathol., 23(2), 203-213.

Adamo, N.J. & Ratner, A. (1970). Science, 169, 673-674.

Adkins, J.S., Wertz, J.M., Boffmann, R.H., & Hove, E.L. (1967). Proc. Soc. Exp. Biol. (NY), 126, 500-504.

Airoldi, L., Bizzi, A., Salmona, M., & Garattini, S. (1979a). Attempts to establish the safety margin for neurotoxicity of monosodium glutamate. Glutamic acid (ed. by Filer, L.J. et al.). Raven Press, New York, USA, 321–331.

Airoldi, L., Salmona, M., Ghezzi, P., & Garattini, S. (1979b). Glutamic acid and sodium levels in the nucleus arcuatus of the hypothalamus of adult and infant rats after oral monosodium glutamate. Toxicol. Lett., 3, 121–126.

Ajinomoto Co. (1970). Unpublished report. Submitted to WHO by Ajinomoto, Co., Inc.

Aleksandrov, P.N., Bogdanova, V.A., & Chernukh, A.M. (1965). Farm. i. Toks., 28(6), 744.

Anantharaman, K. (1979). In utero and dietary administration of monosodium L-glutamate to mice: reproductive performance and development in a multigeneration study (ed. by Filer, L.J. et al). Raven Press, New York, USA, 231–253.

Araujo, P.E. & Mayer, J. (1973). Activity increase associated with obesity induced by monosodium glutamate in mice. Amer. J. Physiol., 225(4), 764–765.

Arees, E. & Mayer, J. (1971). Paper presented at the 47th Annual Meeting of Amer. Assoc. Neuropath. Puerto Rico, June.

Baker, G.L., Filer, L.J., & Stegink, L.D. (1979). Factors influencing dicarboxylic amino acid content of human milk. Glutamic acid (ed. by Filer, L.J. et al.). Raven Press, New York, USA, 111–124.

Berfenstam, R., Jazenburg, R., & Mellander, O. (1955). Acta Paed., 44, 348.

Beron, E.L. (1968). Letter to the editor. New Engl. J. Med., 278, 1123.

Berry, W.T.C. (1970). Unpublished material from surveys. Department of Health and Social Security, London, England.

Berry, H.K., Butcher, R.E., Elliot, L.A., & Brunner, R.L. (1974). The effect of monosodium glutamate on the early biochemical and behavioural development of the rat. Develop. Psychobiol., 7, 165–173.

Bhagavan, H.N., Coursin, D.B., & Stewart, C.N. (1971). Nature, 232, 275–276.

Bizzi, A., Veneroni, E., Salmona, M., & Garattini, S. (1977). Kinetics of monosodium glutamate in relation to its neurotoxicity. Toxicol. Lett., 1, 123–130.

Boaz, D.P., Stegink, L.D., Reynolds, W.A., Filer, L.J. Jr., Pitkin, R.M., & Brummel, M.C. (1974). Monosodium glutamate metabolism in the neonatal primate. Fed. Proc., 33, 651.

Brand, J.G., Cagean, R.H., & Naim, M. (1982). Chemical senses in the release of gastric and pancreatic secretions. Ann. Rev. Nutr., 2, 249-276.

Burde, R.M., Schainker, B., & Kayes, J. (1971). Acute effect of oral and subcutaneous administration of monosodium glutamate on the arcuate nucleus of the hypothalamus in mice and rats. Nature, 233, 58-60.

Byun, S.M. (1980). Human development: human serum glutamate levels of Koreans. Nutr. Rep. Int., 22(5), 697-705.

Caccia, S., Garattini, S., Ghezzi, P., & Zanini, M.G. (1982). Plasma and brain levels of glutamate and pyroglutamate after oral monosodium glutamate to rats. Toxicol. Lett., 10(2-3), 169-175.

Clemens, J.A., Roush, M.E., Fuller, R.W., & Shaar, C.J. (1978). Changes in luteinizing hormone and prolactin control mechanisms produced by glutamate lesions of arcuate nucleus. Endocrin., 103, 1304-1312.

Cohen, P.P. (1949). Biochem. J., 33, 1478.

Cohen, A.I. (1967). An electron microscopic study of the modification by monosodium glutamate of the retinas of normal and "rodless" mice. Amer. J. Anat., 120, 319-356.

Crawford, J.M. (1963). J. Biol. Chem., 240, 1443.

Daabees, T.T., Andersen, D.W., Zike, W.L., Filer, L.J., & Stegink, L.D. (1984). Effect of meal components on peripheral and portal plasma glutamate levels in young pigs administered large doses of monosodium L-glutamate. Metabolism, 33, 58-67.

Daabees, T.T., Fikelstein, M.W., Stegink, L.D., & Applebaum, A.E. (1985). Correlation of glutamate plus aspartate dose, plasma amino acid concentration and neuronal encrosis in infant mice. Food Chem. Toxic., 23(10), 887-893.

Daniel, R.G. & Waisman, H.A. (1968). Growth, 32, 255.

Dhindsa, K.S., Omran, R.G., & Bhup, R. (1978). Effect of monosodium glutamate on the histogenesis of bone and bone marrow in mice. Acta Anat., 101, 212-217.

Dickinson, J.C. & Hamilton, P.B. (1966). The free amino acids of human spinal fluid determined by ion exchange chromatography. J. Neurochem., 13, 1179-1187.

Ebert, A.G. (1970). Chronic toxicity and teratology studies of L-monosodium glutamate and related compounds. Toxicol. Appl. Pharmacol., 17, 274.

Ebert, A.G. (1979a). The dietary administration of monosodium glutamate or glutamic acid to C-57 black mice for two years. Toxicol. Lett., 3, 65-70.

Ebert, A.G. (1979b). The dietary administration of L-monosodium glutamate, DL-monosodium glutamate, or L-glutamic acid to rats. Toxicol. Lett., 3, 71-78.

von Euler, H., Adler, E., Gunther, G., & Das, N.B. (1938). Z. Physiol. Chem., 254, 61.

Everly, J.L. (1971). Light microscopy examination of monosodium glutamate-induced lesions in the brain of fetal and neonatal rats. Anat. Rec., 169, 312.

Fernandez-Flores, E., Kline, D.A., Johnson, A.R., & Leber, B.L. (1970). Quantitative and qualitative GLC analysis of free amino acids in fruits and fruit juices. J. AOAC, 53, 1203-1208.

Food and Drug Research Laboratories (1974a). Teratologic evaluation of FDA 71-69 (monosodium glutamate) in mice, rats and rabbits. NTIS, PB-234-865.

Food and Drug Research Laboratories (1974b). Teratologic evaluation of FDA 73-58 (monopotassium glutamate) in mice and rats. Unpublished report, November 27.

Francesconi, R.P. & Villee, C.A. (1968). Biochem. Biophys. Res. Comm., 31, 713.

Freedman, J.K. & Potts, A.M. (1962). Invest. Ophthal., 1, 118.

Freedman, J.K. & Potts, A.M. (1963). Invest. Ophthal., 2, 252.

Furuya, H. (1967). Unpublished report. Submitted to WHO in 1970.

Garattini, S. (1971). Unpublished report from Istituto di Ricerche Farmacologiche "Mario Negri", 20157 Milan, Italy. Submitted to WHO in 1973.

Geil, R.G. (1970). Prelim. Comm. Gerber Products Co. Submitted to WHO in 1973.

Ghadimi, H., Kumar, S., & Abaci, F. (1971). Studies on monosodium glutamate ingestion. I. Biochemical explanation of Chinese restaurant syndrome. Biochem. Med., 5, 447-456.

Ghezzi, P., Salmona, M., Recchia, M., Dagnino, G., & Garattini, S. (1980). Monosodium glutamate kinetic studies in human volunteers. Toxicol. Lett., 5, 417-421.

Ghezzi, P., Bianchi, M., Gianera, L., Salmona, M., & Garattini, S. (1985). Kinetics of monosodium glutamate in human volunteers under different experimental conditions. Food Chem. Toxicol., 23(11), 975-978.

Giacometti, T. (1979). Free and bound glutamate in natural products. Glutamic acid (ed. by Filer, L.J. et al.). Raven Press, New York, USA, 25-34.

Go, G., Nakamura, F.H., Rhoads, G.G., & Dickinson, L.E. (1973). Long-term health effects of dietary monosodium glutamate. Hawaii Medical J., 32, 13-17.

Golberg, L. (1973). Unpublished report from the Institute of Experimental Pathology and Toxicology, Albany Medical College. Submitted to WHO in 1973.

Gottschewski, G.H.M. (1967). Kann die Traegersubstanz von Wirkstoffen in Dragees eine teratogene Wirkung haben? Arzneimittel-Forsch., 17, 1100-1103.

GRAS (1976). Committee on GRAS list survey - phase III: Estimating distributions of daily intake of monosodium glutamate (MSG), Appendix E, in Estimating Distribution of Daily Intake of Certain GRAS substances, Food and Nutrition Board, Division of Biological Sciences, Assembly of Life Sciences, NRC/NAS, Washington, DC, USA.

Hamatsu, T. (1964). Experimental studies on the effect of sodium iodate and sodium L-glutamate on ERG and histological structure of retina in adult rabbits. Acta Soc. Ophthalmol. Jpn., 68, 1621-1636.

Hansson, H.A. (1970). Ultrastructural studies on the long-term effects of sodium glutamate on the rat retina. Virchows Arch. B., 6, 1-11.

Hara, S., Shibuya, T., Nakakawaji, K., Kyu, M., Nakamura, Y., Hoshikawa, H., Takeuchi, T., Iwao, T., & Ino, H. (1962). Observations of pharmacological actions and toxicity of sodium glutamate, with comparisons between natural and synthetic products. Tokyo Idadaigaku Zasshi, J. Tokyo Med. Coll., 20(1), 69-100.

Hazelton Laboratories (1966). Report to International Mineral and Chemical Corporation dated 3 November 1966. Submitted to WHO in 1970.

Hazelton Laboratories (1969a). Addendum to report of 1966 dated 18 July 1969. Submitted to WHO in 1970.

Hazelton Laboratories (1969b). Addendum to report of 1966 dated 10 December 1969. Submitted to WHO in 1970.

Hepburn, F.N., Calhoun, W.K., & Bradley, W.B. (1960). J. Nutr., 72, 163.

Hepburn, F.N. & Bradley, W.B. (1964). J. Nutr., 84, 305.

Herbst, A., Wiechert, P., & Hennecke, H. (1966). Experientia, 22(11), 718.

Heywood, R., Palmer, A.K., & Hague, P.H. (1971). Unpublished report submitted to Central Research Laboratories, Ajinomoto Co., Inc. Submitted to WHO in 1973.

Heywood, R., Palmer, A.K., & Newman, A.J. (1972a). Unpublished report submitted to Central Research Laboratories, Ajinomoto Co., Inc. Submitted to WHO in 1973.

Heywood, R., Palmer, A.K., Newman, A.J., Barry, D.H. & Edwards, F.P. (1972b). Unpublished report submitted to Central Research Laboratories, Ajinomoto Co., Inc. Submitted to WHO in 1973.

Heywood, R., James, R.W., & Worden, A.N. (1977). The ad libitum feeding of monosodium glutamate to weanling mice. Toxicol. Lett., 1, 151-155.

Heywood, R., James, R.W., & Salmona, M. (1978). Serum glutamate in rhesus macaques after gavage with MSG. Toxicol. Lett., 2, 299-303.

Heywood, R. & James, R.W. (1979). An attempt to induce neurotoxicity in an infant rhesus monkey with monosodium glutamate. Toxicol. Lett., 4, 285-286.

Heywood, R. & Worden, A.N. (1979). Glutamate toxicity in laboratory animals. Glutamic acid (ed. by Filer, L.J. et al.). Raven Press, New York, USA, 203-215.

Himwich, W.A. (1954). Science, 120, 351.

Himwich, H.E., Wolff, K., Hunsicker, A.L., & Himwich, W.A. (1954). Appl. Physiol., 7, 40.

Holzwarth-McBride, M.A., Sladek, J.R., & Knigge, K.M. (1976). Monosodium glutamate-induced lesions of the arcuate nucleus. II. Fluorescence histochemistry of catecholamines. Anat. Rec., 186(2), 197-205.

Huang, P.C., Lee, N.Y., Wu, T.J., Yu, S.L., & Tung, T.C. (1976). Effect of monosodium glutamate supplementation to low protein diets on rats. Nutr. Rep. Int., 13, 477-486.

Huntington Research Centre (1971). Unpublished report dated 18 October. Submitted to WHO in 1973.

Ichimura, M. & Kirimura, J. (1968). Unpublished report from Central Research Laboratories, Ajinomoto Co., Inc. Submitted to WHO in 1970.

Ikeda, K. (1908). Method of producing a seasoning material whose main component is the salt of glutamic acid. Japanese Pat. No. 14, 805.

Ikeda, K. (1909). On the taste of the salt of glutamic acid. J. Tokyo Chem. Soc., 30, 820-836.

Ikeda, K. (1912). On the taste of the salt of glutamic acid. Eighth Int. Cong. Appl. Chem., 147.

Industrial Bio-test Laboratories (1973a). Mutagenic study with accent brand monosodium L-glutamate in albino mice. Northbrook, IL, USA, 1-12.

Industrial Bio-test Laboratories (1973b). Host-mediated assay for detection of mutations induced by accent brand monosodium L-glutamate. Northbrook, IL, USA, 1-19.

International Min. and Chem. Corp. (1969). Unpublished report. Submitted to WHO by International Mineral and Chemical Corporation.

Iwata, S., Ichimura, M., Matsuzawa, Y., Takasaki, Y., & Sasaoka, M. (1979). Behavioural studies in rats treated with monosodium L-glutamate during the early stages of life. Toxicol. Lett., 4(5), 345-357.

Izeki, T. (1964). Report of the Osaka Municipal Hygienic Laboratory, 23, 82. Submitted to WHO in 1970.

James, R.W., Heywood, R., Worden, A.N., Garattini, S., & Salmona, M. (1978). The oral administration of MSG at varying concentrations to male mice. Toxicol. Lett., 1, 195-199.

Kandall, S.R. (1968). Letter to the editor. New Engl. J. Med., 278, 1123.

Kawamura, Y. & Kare, M.R. (1987). Umami: a basic taste. Marcel Dekker, New York and Basel.

Kenney, R.A. (1986). The Chinese restaurant syndrome: an anecdote revisited. Food Chem. Toxicol., 24(4), 351-354.

Kerr, G.R., Wu-Lee, M., El-Lozy, M., McGandy, R., & Stare, F.J. (1977). Objectivity of food-symptomatology surveys. Questionnaire on the Chinese restaurant syndrome. J. Am. Diet Assoc., 71(3), 263-268.

Kerr, G.R., Wu-Lee, M., El-Lozy, M., McGandy, R., & Stare, F.J. (1979a). Food symptomatology questionnaires: risks of demand bias questions and population biased surveys. Glutamic acid (ed. by Filer, L.J. et al.). Raven Press, New York, USA, 375-387.

Kerr, G.R., Wu-Lee, M., El-Lozy, M., McGandy, R., & Stare, F.J. (1979b). Prevalence of the "Chinese restaurant syndrome". J. Am. Diet Assoc., 75(1), 29-33.

Khera, K.S., Whitta, L.L., & Nera, E.A. (1970). Unpublished results of Research Laboratories, Food and Drug Directorate, Ottawa, Canada. Submitted to WHO in 1970.

Kirimura, J., Shimizu, A., Kimizuka, A., Nimomiya, T., & Katsuya, N. (1969). The contribution of peptides and amino acids to the taste of foodstuffs. J. Agric. Food Chem., 17, 689-695.

Krnjevic, K. (1970). Nature, 228, 119-124.

Kwok, R.H.M. (1968). Chinese-restaurant syndrome. New Engl. J. Med., 278, 796.

Lamperti, A. & Blaha, G. (1976). The effects of neonatally-administered monosodium glutamate on the reproductive system of adult hamsters. Biol. Reprod., 14, 362-369.

Lamperti, A. & Blaha, G. (1980). Further observations on the effects of neonatally-administered monosodium glutamate on the reproductive axis of hamsters. Biol. Reprod., 22, 687-693.

Lechan, R.M., Alpert, L.C., & Jackson, I.M. (1976). Synthesis of luteinising hormone releasing factor and thyrotrophin-releasing factor in glutamate-lesioned mice. Nature., 264(5585), 463-465.

Lemkey-Johnston, N. & Reynolds, W.A. (1972). Incidence and extent of brain lesions in mice following ingestion of monosodium glutamate (MSG). Anat. Rec., 172, 354.

Lemkey-Johnston, N., & Reynolds, W.A. (1974). Nature and extent of brain lesions in mice related to ingestion of monosodium glutamate. A light and electron microscopy study. J. Neuropathol. Exp. Neurol., 33(1), 74-97.

Lemkey-Johnston, N., Butler, V., & Reynolds, W.A. (1975). Brain damage in neonatal mice following monosodium glutamate administration: possible involvement of hypernatremia and hyperosmolality. J. Neuropathol. Exp. Neurol., 48(2), 292-309.

Lengvari, T. (1977). Effect of perinatal monosodium glutamate treatment on endocrine functions of rats in maturity. Acta. Biol. Acad. Sci. Hung., 28, 133-141.

Levey, S., Harroun, J.E., & Smyth, C.J. (1949). Serum glutamic acid levels and the occurrence of nausea and vomiting after the intravenous administration of amino acid mixtures. J. Lab. Clin. Med., 34, 1238-1248.

Liebschultz, J., Airoldi, L., Brounstein, M.J., & Chinn, N.G. (1977). Regional distribution of endogenous and parenteral glutamate, aspartate, and glutamine in rat brain. Biochem. Pharmacol., 26, 443-446.

Little, A.D. (1953a). Report submitted to International Mineral and Chemical Corporation dated 13 January 1953. Submitted to WHO in 1970.

Little, A.D. (1953b). Report submitted to International Mineral and Chemical Corporation dated 15 March 1953. Submitted to WHO in 1970.

Litton Bionetics (1975a). Mutagenic evaluation of compound FDA 73-58, 000997-42-2, monopotassium glutamate. US Department of Commerce, National Technical Information Service PB-254.511.

Litton Bionetics (1975b). Mutagenic evaluation of compound FDA 75-11, 007558-63-6, monoammonium glutamate, FCC. US Department of Commerce, National Technical Information Service PB-254.512.

Litton Bionetics (1977a). Mutagenic evaluation of compound FDA 75-59, L-glutamic acid •HCl. US Department of Commerce, National Technical Information Service PB-266.892.

Litton Bionetics (1977b). Mutagenic evaluation of compound FDA 75-65, L-glutamic acid, FCC. US Department of Commerce, National Technical Information Service PB-266.889.

Lucas, D.R. & Newhouse, J.P. (1957). The toxic effect of sodium L-glutamic acid on the inner layers of the retina. Amer. Med. Assoc. Arch. Ophthalm., 5, 193-201.

Madden, S.C., Woods, R.R., Skull, F.W., & Whipple, G.H. (1944). J. Exp. Med., 79, 607.

Maeda, S., Eguchi, S., & Sasaki, H. (1958). The content of free L-glutamic acid in various foods. J. Home Econ., 9, 163-167.

Maeda, S., Eguchi, S., & Sasaki, H. (1961). The content of free L-glutamic acid in various foods (part 2). J. Home Econ., 12, 105-106.

Maga, J.A. (1983). Flavor Potentiators. CRC Critical Reviews in Food Science and Nutrition, 18, 231-312.

Marrs, T.C., Salmona, M., Garattini, S., Murston, D., & Matthews, D.M. (1978). The absorption by human volunteers of glutamic acid from monosodium glutamate and from a partial enzymic hydrolysate of casein. Toxicology, 11, 101-107.

Martindale, L.W. (1967). Extra Pharmacopoeia, 25th ed.

Matsuyama, S., Oki, Y., & Yokoki, Y. (1973). Obesity induced by monosodium glutamate in mice. Nat. Inst. Anim. Health, 13, 91-101.

Matsuzawa, Y., Yonetani, S., Takasaki, Y., Iwata, S., & Sekiny, S. (1979). Studies on reproductive endocrine function in rats treated with monosodium L-glutamate early in life. Toxicol. Lett., 4, 359-371.

Matthews, D.M. (1975). Intestinal absorption of peptides. Physiol. Rev., 55, 537-608.

Matthews, D.M. (1984). Absorption of peptides, amino acids, and their methylated derivatives. Aspartame, Physiology and Biochemistry, edited by L.D. Stegink et al., pp. 29-46, Marcel Dekker, New York, USA.

Mazurowa, A., Mrozikiewicz, A., & Wrocinski, T. (1962). Acta Physiol. Polonica, 13, 797.

McColl, J.D., Globus, M., & Robinson, S. (1965). Canad. J. Phys. Pharm., 43, 69.

Meister, A. (1965). Biochemistry of the amino acids. Vol I and II, 2nd ed. Academic Press.

Meister, A. (1979). Biochemistry of glutamate: glutamine and glutathione. Glutamic acid (ed. by Filer, L.J. et al.). Raven Press, New York, USA, 69-84.

Moriyuki, H. & Ichimura, M. (1978). Acute toxicity of monosodium L-glutamate in mice and rats. Oyo Yakuri., 15, 433-436.

Mueller, H. (1970). Ueber das Vorkommen freier glutaminsaure in lebensmitteln. Ernaehrungswissenschaft, 10, 83-88.

Munro, H.N. (1979). Factors in the regulation of glutamate metabolism in glutamic acid (ed. by Filer, L.J. et al.). Raven Press, New York, USA, 55-68.

Murakami, U. & Inouye, M. (1971). Brain lesions in the mouse fetus caused by maternal administration of monosodium glutamate (preliminary report). Cong. Anom., 11, 171-177.

Mushahwar, I.K. & Koeppe, R.E. (1971). Biochim. Biophys. Acta, 244, 318-321.

Nagasawa, H., Yanai, R., & Kikuyama, S. (1974). Irreversible inhibition of pituitary prolectin and growth hormone secretion and of mammary gland development in mice by monosodium glutamate administered neonatally. Acta Endocrinol., 75, 249-259.

Nemeroff, C.B., Ervin, G.N., Bissette, G., Harrell, L.E., & Grant, L.D. (1975). Growth and endocrine alterations after neonatal monosodium L-glutamate in the rat: evaluation of relationship to arcuate dopamine neuron damage. Toxicol. Appl. Pharmacol., 33, 163.

Nemeroff, C.B., Bissette, G., & Kizer, J.S. (1977a). Reduced levels of immunoreactive LHRH in genetically obese mice (ob/ob). The Proceedings of Soc. Neurosci. Symp., 7th Annual Meeting, 353.

Nemeroff, C.B., Konkol, R.J., Bissette, G., Youngblood, W., Marin, J.B., Brazeau, P., Rone, M.S., Prange, A.J. Jr., Breese, G.R., & Kizer, J.S. (1977b). Analysis of the disruption in hypothalamic-pituitary regulation in rats treated neonatally with monosodium L-glutamate (MSG). Evidence for the involvement of tuberoinfundibular cholinergic and dopaminergic systems in neuroendocrine regulation. Endocrinology, 101, 613-622.

Nemeroff, C.B., Grant, L.D., Bissette, G., Ervin, G.N., Harrell, L.E., & Prange, A.J. Jr. (1977c). Growth, endocrinological and behavioural deficits after monosodium L-glutamate in the neonatal rat. Possible involvement of arcuate dopamine neuron damage. Psychoneuroendocrinology, 2, 179-196.

Newman, A.J., Heywood, R., Palmer, A.K., Barry, D.H., Edwards, F.P., & Worden, A.N. (1973). The administration of monosodium L-glutamate to neonatal and pregnant rhesus monkeys. Toxicology, 1, 197-204.

Nikoletseas, M.M. (1977). Obesity in exercising hypophagic rats treated with monosodium glutamate. Physiol. Behav., 19, 767-773.

Ohara, Y., Fukimoto, T., Ichimura, M., & Kirimura, J. (1970). Unpublished report from Central Research Laboratories, Ajinomoto Co., Inc. Submitted to WHO in 1970.

Ohara, Y., Iwata, S., Ichimura, M., & Sasaoka, M. (1977). Effect of administration routes of monosodium glutamate on plasma glutamate levels in infant, weanling and adult mice. J. Toxicol. Sci., 2, 281-290.

Oldendorf, W.H. (1971). Brain uptake of radiolabeled amino acids, amines and hexoses after arterial injection. Am. J. Physiol., 221, 1629-1639.

Olivo, M., Kitahama, K., Valatx, J.L., & Jouvet, M. (1986). Neonatal monosodium L-glutamate dosing alters the sleep-wake cycle of the mature rat. Neurosci. L., 67, 186-190.

Olney, J.W. (1969a). Glutamate-induced retinal degeneration in neonatal mice. Electron microscopy of the acutely evolving lesion. J. Neuropath. Exp. Neurol., 28, 455-474.

Olney, J.W. (1969b). Brain lesions, obesity, and other disturbances in mice treated with monosodium glutamate. Science, 164, 719-721.

Olney, J.W. & Sharp, L.G. (1969). Science, 166, 386.

Olney, J.W. (1970). Unpublished report. Submitted to WHO in 1970.

Olney, J.W. & Ho, O.L. (1970). Brain damage in infant mice following oral intake of glutamate, aspartate or cysteine. Nature, 227, 609-611.

Olney, J.W. (1971). Glutamate-induced neuronal necrosis in the infant mouse hypothalamus. J. Neuropath. Exp. Neurol., 30, 75-90.

Olney, J.W., Ho, O.L., & Rhee, V. (1971). Exp. Brain. Res., 14, 61.

Olney, J.W., Sharpe, L.G., & Fergin, R.D. (1972). Glutamate-induced brain damage in infant primates. J. Neuropath. Exp. Neurol., 31, 464-488.

Olney, J.W., Rhee, V., & de Gubareff, T. (1977). Neurotoxic effects of glutamate on mouse area postrema. Brain Res., 120, 151-157.

Olney, J.W., Labruyère, J., & de Gubareff, T. (1980). Brain damage in mice from voluntary ingestion of glutamate and aspartate. Neurobehav. Toxicol., 2, 125-129.

Oser, B.L., Carson, S., Vogin, E.E., & Cox, G.E. (1971). Nature, 229, 411-413.

Oser, B.L., Bailey, D.E., Morgareidge, K., Carson, S., & Vogin, E.E. (1973). Unpublished report submitted to International Glutamate Technical Committee. Submitted to WHO in 1973.

Owen, G., Cherry, C.P., Prentice, D.E., & Worden, A.N. (1978a). The feeding of diets containing up to 4% monosodium glutamate to rats for 2 years. Toxicol. Lett., 1, 221-226.

Owen, G., Cherry, C.P., Prentice, D.E., & Worden, A.N. (1978b). The feeding of diets containing up to 10% monosodium glutamate to beagle dogs for 2 years. Toxicol. Lett. (Amst.), 1, 217-219.

Pagliari, A.S. & Goodman, A.D. (1969). New Eng. J. Med., 281, 767.

Pardridge, W.M. (1979). Regulation of amino acid availability to brain: selective control mechanisms for glutamate. Glutamic acid (ed. by Filer, L.J. et al.). Raven Press, New York, USA, 125-137.

Peng, Y., Gubin, J., Harper, A.E., Vavich, M.G., & Kemmerer, A.R. (1973). Food intake regulation: amino acid toxicity and changes in rat brain and plasma amino acids. J. Nutr., 101, 608-617.

Perez, V.J. & Olney, J.W. (1972). Accumulation of glutamic acid in the arcuate nucleus of the hypothalamus of the infant mouse following subcutaneous administration of monosodium glutamate. J. Neurochem., 19, 1777-1782.

Perrault, M. & Dry, J. (1961). Sem. Thérapeutique, 37, 597.

Peters, J.H., Lin. S.C., Berridge, B.J. Jr., Cummings, J.G., & Chao, W.R. (1969). Amino acids, including asparagine and glutamine, in plasma and urine of normal human subjects. Proc. Soc. Expt. Biol. Med., 131, 281-288.

Pinto-Scognamiglio, W., Amorico, L., & Gatti, G.L. (1972). Esperienze di tossicita et di tolleranza al monosodioglutammato con un saggio di condizionamento di salvaguardia. Il Farmaco (Prat), 27, 19-27.

Pitkin, R.M., Reynolds, W.A., Stegink, L.D., & Filer, L.J. (1979). Glutamate metabolism and placental transfer in pregnancy. Glutamic acid (ed. by Filer, L.J. et al.). Raven Press, New York, USA, 103-110.

Pizzi, W.J., Barnhart, J.E., & Fanslow, D.J. (1977). Monosodium glutamate administration to the newborn reduces reproductive ability in female and male mice. Science, 196, 452-454.

Poon, T.K.-Y. & Cameron, D.P. (1978). Measurement of oxygen consumption and locomotor activity in monosodium glutamate-induced obesity. Am. J. Physiol., 234, 532-534.

Porter, D.B. & Griffin, A.C. (1950). J. Comp. Phys. Psych., 43, 1.

Potts, A.M., Modrell, R.W., & Kingsbury, C. (1960). Permanent fractionation of the electroretinogram by sodium glutamate. Amer. J. Ophthal., 50, 900-907.

Prabhu, V.G. & Oester, Y.T. (1971). Neuromuscular functions of mature mice following neonatal monosodium glutamate. Arch. Int. Pharmacodyn. Ther., 189, 59-71.

Pradhan, S.N. & Lynch, J.F. Jr. (1972). Behavioural changes in adult rats treated with monosodium glutamate in the neonatal stage. Arch. Int. Pharmacodyn. Ther., 197, 301-304.

Price, J.C., Waelsch, H., & Putmann, L.F. (1943). J. Amer. Med. Assoc., 122, 1153.

Price, M.T., Olney, J.W., & Lowry, O.H. (1981). Uptake of exogenous glutamate and aspartate by circumventricular organs but not other regions of brain. J. Neurochem., 36, 1774-1780.

Prosky, L. & O'Dell, R.G. (1972). Biochemical changes of brain and liver in neonatal offspring of rats fed monosodium L-glutamate. Experientia, 28, 260-263.

Rassin, D.K., Sturman, J.A., & Caull, G.E. (1978). Taurine and other free amino acids in milk of man and other mammals. Early Human Dev., 2, 1-13.

Redding, T.W., Schally, A.V., Arimura, L.A., & Wakabayashi, I. (1971). Effect of monosodium glutamate on some endocrine functions. Neuroendocrinology, 8, 245-255.

Reynolds, W.A., Lemkey-Johnston, N., Filer, L.J. Jr., & Pitkin, R.M. (1971). Monosodium glutamate: absence of hypothalamic lesions after ingestion by newborn primates. Science, 172, 1342-1344.

Reynolds, W.A., & Lemkey-Johnston, N. (1973). Unpublished data submitted to International Glutamate Technical Committee. Submitted to WHO in 1973.

Reynolds, W.A., Butler, V., & Lemkey-Johnston, N. (1976). Hypothalamic morphology following ingestion of aspartame or MSG in the neonatal rodent and primate: a preliminary report. J. Toxicol. Environ. Health., 2, 471-480.

Reynolds, W.A., Lemkey-Johnston, N., & Stegink, L.D. (1979). Morphology of the fetal monkey hypothalamus after in utero exposure to monosodium L-glutamate. Glutamic acid (ed. by Filer L.J. et al.). Raven Press, New York, USA, 217-229.

Robinzon, B., Snapir, N., & Perek, M. (1974). Age-dependent sensitivity to monosodium glutamate inducing brain damage in the chicken. Poult. Sci., 53, 1539-1542.

Robinzon, B., Snapir, N., & Perek, M. (1975). The relation between monosodium glutamate inducing brain damage, and body weight, food intake, semen production, and endocrine criteria in the fowl. Poult. Sci., 54, 234-241.

Rogers, L.J. (1982). Teratological effects of glutamate on behaviour. Food Technol. in Australia, 34, 202-206.

Sauberlich, H.E. (1961). J. Nutr., 75, 61.

Schaumburg, H.H., Byck, R., Gerstl, R., & Mashman, J.H. (1969). Monosodium L-glutamate: its pharmacology and role in the Chinese restaurant syndrome. Science, 163, 826-828.

Schneider, H., Moehlenkii, Challier, J.C., & Danicis, J. (1979).
Transfer of glutamic acid across the human placenta perfused in vitro.
Br. J. Obstet. Gynecol., 86, 299-306.

Schultz, S.G., Yu-Tu, L., Alvarez, O.O., & Curran, P.F. (1970).
Dicarboxylic amino acid influx across brush border of rabbit ileum. J.
Gen. Physiol., 56, 621-639.

Semprini, M.E., Frasca, M.A., & Mariani, A. (1971). Effects of
monosodium glutamate (MSG) administration on rats during the
intrauterine life and the neonatal period. Quaderni delle Nutrizione,
31, 85-100.

Semprini, M.E., Conti, L., Ciofi-Luzzatto, A., & Mariani, A. (1974).
Effect of oral administration of monosodium glutamate (MSG) on the
hypothalamic arcuate region of rat and mouse: a histological assay.
Biomedicine, 21, 398-403.

Shimizu, T. & Aibara, K. (1970). Unpublished report. Submitted to WHO
in 1970.

Snapir, N., Robinzon, B., & Perek, M. (1971). Brain damage in the male
domestic fowl treated with monosodium glutamate. Poult. Sci., 50,
1511-1514.

Solms, J. (1969). The taste of amino acids, peptides, and proteins.
J. Agr. Food Chem., 17, 686-688.

Stegink, L.D., Filer, L.J. Jr., & Baker, G.L. (1972). Monosodium
glutamate: effect on plasma and breast milk amino acid levels in
lactating women. Proc. Soc. Expt. Biol. & Med., 140, 836-841.

Stegink, L.D., Pitkin, R.M., Reynolds, W.A., & Filer, L.J. Jr. (1973).
Federation Proceedings, 32, Abstract 3797.

Stegink, L.D., Shepherd, J.A., Brummel, M.C., & Murray, L.M. (1974).
Toxicity of protein hydrolysate solutions: correlation of glutamate
dose and neuronal necrosis to plasma amino acid levels in young mice.
Toxicology, 2, 285-299.

Stegink, L.D., Pitkin, R.M., Reynolds, W.A., Filer, L.J., Boaz, D.P., &
Brummel, M.C. (1975a). Placental transfer of glutamate and its
metabolites in the primate. Am. J. Bostet. Gynecol., 122, 70-78.

Stegink, L.D., Reynolds, W.A., Filer, L.J., Pitkin, R.M., Boaz, D.P., &
Brummel, M.C. (1975b). Monosodium glutamate metabolism in the neonatal
monkey. Am. J. Physiol., 122, 70-78.

Stegink, L.D., Reynolds, W.A., Filer, L.J., Baker, G.L., Daabees, T.T.,
& Pitkin, R.M. (1979a). Comparative metabolism of glutamate in the
mouse, monkey and man. Glutamic acid (ed. by Filer, L.J. et al).
Raven Press, New York, USA, 85-102.

Stegink, L.D., Filer, L.J., Baker, G.L., Mueller, S.M., & Wu-Rideout,
M. (1979b). Factors affecting plasma glutamate levels in normal adult

subjects. Glutamic acid, advances in biochemistry and physiology (ed. by Filer, L.J. et al). Raven Press, New York, USA, 33-351.

Stegink, L.D., Filer, L.J. Jr., & Baker, G.L. (1982). Plasma and erythrocyte amino acid levels in normal adults subjects fed a high protein meal with and without added monosodium glutamate. J. Nutr., 112, 1953-1960.

Stegink, L.D., Baker, G.L., & Filer, L.J. (1983a). Modulating effect of sustagen on plasma glutamate concentration in humans ingesting monosodium L-glutamate. Am. J. Clin. Nutr., 194-200.

Stegink, L.D., Filer, L.J., & Baker, G.L. (1983b). Effect of carbohydrate on plasma and erythrocyte glutamate levels in humans ingesting large doses of monosodium L-glutamate in water. Am. J. Clin. Nutr., 37, 961-968.

Stegink, L.D., Filer, L.J. Jr., & Baker, G.L. (1983c). Plasma amino acid concentrations in normal adults fed meals with added monosodium L-glutamate and aspartame. J. Nutr., 113, 1851-1860.

Stegink, L.D. (1984). Aspartate and glutamate metabolism. Aspartame: physiology and biochemistry. pp. 47-76.

Stegink, L.D., Filer, L.J., & Baker, G.L. (1985a). Effect of starch ingestion on plasma glutamate concentrations in humans ingesting monosodium L-glutamate in soup. J. Nutr., 115, 211-218.

Stegink, L.D., Filer, L.J., & Baker, G.L. (1985b). Plasma glutamate concentrations in adult subjects ingesting monosodium L-glutamate in consomme. Am. J. Clin. Nutr., 42, 220-225.

Stegink, L.D., Filer, L.J., Baker, G.L., & Bell, E.F. (1986). Effect of sucrose ingestion on plasma glutamate concentrations in humans administered monosodium L-glutamate. Am. J. Clin. Nutr., 42, 220-225.

Stellar, E. & McElroy, W.D. (1948). Science, 108, 281.

Sugahara, M. & Ariyoshi, L.S. (1967). Agr. Biol. Chem., 31, 1270-1275.

Suzuki, Y. & Takahashi, M. (1970). Unpublished report. Submitted to WHO by Ajinomoto Co., Inc.

Tafelski, T.J. & Lamperti, A.A. (1977). The effects of a single injection of monosodium glutamate on the reproductive neuroendocrine axis of the female hamster. Biol. Reprod., 17, 404-411.

Takasaki, Y. (1978a). Studies on brain lesions after administration of monosodium L-glutamate to mice. I. Brain lesions in infant mice caused by administration of monosodium L-glutamate. Toxicology, 9, 293-305.

Takasaki, Y. (1978b). Studies on brain lesions after administration of monosodium L-glutamate to mice. II. Absence of brain damage following administration of monosodium L-glutamate in the diet. Toxicology, 9, 307-318.

Takasaki, Y. (1979). Protective effect of mono- and disaccharides on glutamate-induced brain damage in mice. Toxicol. Lett., 4, 205-210.

Takasaki, Y., Matsuzawa, Y., Iwata, S., O'Hara, Y., Yonetani, S., & Ichimura, M. (1979a). Toxicological studies of monosodium L-glutamate in rodents - relationship between routes of administration and neurotoxicity. Glutamic acid (ed. by Filer, L.J. et al). Raven Press, New York, USA, 255-275.

Takasaki, Y. & Yugari, Y. (1980). Protective effect of arginine, leucine and preinjection of insulin on glutamate neurotoxicity in mice. Toxicol. Lett., 5, 39-44.

Takasaki, Y. & Torii, K. (1983). Effects of water restriction on the development of hypothalamic lesions in weaning rodents given MSG. II. Drinking behaviour and physiological parameters in rats (Rattus norvegicus) and golden hamsters (Mesocricetus auratus). Toxicol. Lett., 16, 195-210.

Terry, L.C., Epelbaum, J., Brazeau, P., & Martin, J.B. (1977). Monosodium glutamate: acute and chronic effects on growth hormone prolactin and somatostatin in the rat. Fed. Proc., 36, 364.

Torii, K. & Takasaki, Y. (1983). Effects of water restriction on the development of hypothalamic lesions in weanling rodents given MSG. I. Drinking behaviour and physiological parameters in mice. Toxicol. Lett., 16, 175-194.

Trentini, G.P., Botticelli, A., & Botticelli, C.S. (1974). Effect of monosodium glutamate on the endocrine glands and on the reproductive function of the rat. Fertil. Steril., 25, 478-483.

Tugrul, S. (1965). Action tératogène de l'acide glutamique. Arch. int. Pharmacodyn., 153, 323-333.

Tung, T.C. & Tung, K.S. (1980). Serum free amino acid levels after oral glutamate intake in infant and adult humans. Nutr. Rep. Int., 22, 431-443.

US FDA (1969). US Food and Drug Administration, Bureau of Science-Bureau of Medicine. Report on monosodium glutamate for review by Food Protection Committee, NAS/NRC, Washington, DC, USA.

US FDA (1975). Investigation of the toxic and teratogenic effects of GRAS substances to the developing chicken embryo. Unpublished report, September 14 from the US Food and Drug Administration.

Waelsch, H. (1949). The Lancet, i, 257.

Walshe, J.M. (1953). The Lancet, i, 1075.

Wen, C.P. & Gershoff, S.N. (1972). Effects of dietary vitamin B_6 on the utilization of monosodium glutamate by rats. J. Nutr., 102, 835-840.

Wen, C.P., Hayes, K.C., & Gershoff, S.N. (1973). Effects of dietary supplementation of monosodium glutamate on infant monkeys, weanling rats and suckling mice. Am. J. Clin. Nutr., 26, 803-813.

Wilkin, J.K. (1986). Does monosodium glutamate cause flushing (or merely "Glutamania")? J. Am. Acad. Dermatol., 15, 225-230.

Wurtman, R.J. (1983). Comments on MSG paper. Food Technology in Australia, 35, 66.

Yamaguchi, S. (1979). The umami taste. Food taste chemistry (ed. by Boudreau, J.C., Washington, DC, USA). Amer. Chem. Soc., 33-51.

Yamaguchi, S. & Kimizuka, A. (1979). Psychometric studies on the taste of monosodium glutamate. Glutamic acid (ed. by Filer, L.J. et al). Raven Press, New York, USA, 35-54.

Yamaguchi, S. & Komata, Y. (1984). Umami taste. Primacy and contribution to the tastes of meat and vegetable stocks. Proceedings of the 18th Japanese Symposium on Taste and Smell, 109-112. Chem. Senses, 10, 137.

Yanagisawa, K., Nakamura, T., Miyata, K., Kameda, T., Kitamura, S., & Ito, K. (1961). Nohon Seirigaku Zasshi. J. Physiol. Soc. Japan, 23, 383-385.

Yonetani, S. (1967). Unpublished report from Central Research Laboratories, Ajinomoto Co., Inc. Submitted to WHO in 1970.

Yonetani, S. Ishii, H., & Kirimura, J. (1970). Unpublished report from Central Research Laboratories, Ajinomoto Co., Inc. Submitted to WHO in 1970.

Yonetani, S. & Matsuzawa, Y. (1978). Effect of monosodium glutamate on serum luteinizing hormone and testosterone in adult male rats. Toxicol. Lett., 1, 207-211.

Yonetani, S. Ishii, H., & Kirimura, J. (1979). Effect of dietary administration of monosodium L-glutamate on growth and reproductive functions in mice. Oyo Yakuri (Pharmacometrics), 17, 143-152.

Zimmerman, F.T. & Burgemeister, B.B. (1959). AMA Arch. Neur. Psych., 81, 639.

ANNEXES

ANNEX 1

REPORTS AND OTHER DOCUMENTS RESULTING FROM MEETINGS OF THE
JOINT FAO/WHO EXPERT COMMITTEE ON FOOD ADDITIVES

1. General principles governing the use of food additives
    (First report of the Joint FAO/WHO Expert Committee on Food
    Additives). FAO Nutrition Meetings Report Series, No. 15,
    1958; WHO Technical Report Series, No. 129, 1957 (out of
    print).

2. Procedures for the testing of intentional food additives to
    establish their safety for use (Second report of the Joint
    FAO/WHO Expert Committee on Food Additives). FAO Nutrition
    Meetings Report Series, No. 17, 1958; WHO Technical Report
    Series, No. 144, 1958 (out of print).

3. Specifications for identity and purity of food additives
    (antimicrobial preservatives and antioxidants) (Third
    report of the Joint FAO/WHO Expert Committee on Food
    Additives). These specifications were subsequently revised
    and published as Specifications for identity and purity of
    food additives, Vol. I. Antimicrobial preservatives and
    antioxidants, Rome, Food and Agriculture Organization of
    the United Nations, 1962 (out of print).

4. Specifications for identity and purity of food additives
    (food colours) (Fourth report of the Joint FAO/WHO Expert
    Committee on Food Additives). These specifications were
    subsequently revised and published as Specifications for
    identity and purity of food additives, Vol. II. Food
    colours, Rome, Food and Agriculture Organization of the
    United Nations, 1963 (out of print).

5. Evaluation of the carcinogenic hazards of food additives
    (Fifth report of the Joint FAO/WHO Expert Committee on Food
    Additives). FAO Nutrition Meetings Report Series, No. 29,
    1961; WHO Technical Report Series, No. 220, 1961 (out of
    print).

6.   **Evaluation of the toxicity of a number of antimicrobials
      and antioxidants** (Sixth report of the Joint FAO/WHO Expert
      Committee on Food Additives). FAO Nutrition Meetings Report
      Series, No. 31, 1962; WHO Technical Report Series, No. 228,
      1962 (out of print).

7.   **Specifications for the identity and purity of food
      additives and their toxicological evaluation; emulsifiers,
      stabilizers, bleaching and maturing agents** (Seventh report
      of the Joint FAO/WHO Expert Committee on Food Additives).
      FAO Nutrition Meetings Report Series, no. 35, 1964; WHO
      Technical Report Series, No. 281, 1964 (out of print).

8.   **Specifications for the identity and purity of food addi-
      tives and their toxicological evaluation: food colours and
      some antimicrobials and antioxidants** (Eighth report of the
      Joint FAO/WHO Expert Committee on Food Additives). FAO
      Nutrition Meetings Report Series, No. 38, 1965; WHO
      Technical Report Series, No. 309, 1965 (out of print).

9.   **Specifications for identity and purity and toxicological
      evaluation of some antimicrobials and antioxidants.** FAO
      Nutrition Meetings Report Series, No. 38A, 1965; WHO/Food/
      Add/24.65 (out of print).

10.   **Specifications for identity and purity and toxicological
      evaluation of food colours.** FAO Nutrition Meetings Report
      Series, No. 35B, 1966; WHO/Food Add/66.25.

11.   **Specifications for the identity and purity of food addi-
      tives and their toxicological evaluation; some anti-
      microbials, antioxidants, emulsifiers, stabilizers, flour-
      treatment agents, acids, and bases** (Ninth report of the
      Joint FAO/WHO Expert Committee on Food Additives). FAO
      Nutrition Meetings Report Series, No. 40, 1966; WHO
      Technical Report Series, No. 339, 1966 (out of print).

12.   **Toxicological evaluation of some antimicrobials, antioxi-
      dants, emulsifiers, stabilizers, flour-treatment agents,
      acids, and bases.** FAO Nutrition Meetings Report Series,
      No. 40A, B, C; WHO/Food Add/67.29.

13.   **Specifications for the identity and purity of food addi-
      tives and their toxicological evaluation; some emulsifiers
      and stabilizers and certain other substances** (Tenth report
      of the Joint FAO/WHO Expert Committee on Food Additives).
      FAO Nutrition Meetings Report Series, No. 43, 1967; WHO
      Technical Report Series, No. 373, 1967.

14.   **Specifications for the identity and purity of food addi-
      tives and their toxicological evaluation; some flavouring
      substances and non-nutritive sweetening agents** (Eleventh
      report of the Joint FAO/WHO Expert Committee on Food
      Additives). FAO Nutrition Meetings Report Series, No. 44,
      1968; WHO Technical Report Series, No. 383, 1968.

15.   Toxicological evaluation of some flavouring substances and
        non-nutritive sweetening agents. FAO Nutrition Meetings
        Report Series, No. 44A, 1968; WHO/Food Add/68.33.

16.   Specifications and criteria for identity and purity of some
        flavouring substances and non-nutritive sweetening agents.
        FAO Nutrition Meetings Report Series, No. 44B, 1969; WHO/
        Food Add/69.31.

17.   Specifications for the identity and purity of food addi-
        tives and their toxicological evaluation; some antibiotics
        (Twelfth report of the Joint FAO/WHO Expert Committee on
        Food Additives). FAO Nutrition Meetings Report Series, No.
        45, 1969; WHO Technical Report Series, No. 430 , 1969.

18.   Specifications for the identity and purity of some anti-
        biotics. FAO Nutrition Meetings Report Series, No. 45A,
        1969; WHO/Food Add/69.34.

19.   Specifications for the identity and purity of food addi-
        tives and their toxicological evaluation; some food
        colours, emulsifiers, stabilizers, anticaking agents, and
        certain other substances (Thirteenth report of the Joint
        FAO/WHO Expert Committee on Food Additives). FAO Nutrition
        Meetings Report Series, No. 46, 1970; WHO Technical Report
        Series, No. 445, 1970.

20.   Toxicological evaluation of some food colours, emulsi-
        fiers, stabilizers, anticaking agents, and certain other
        substances. FAO Nutrition Meetings Report Series, No. 46A,
        1970; WHO/Food Add/70.36.

21.   Specifications for the identity and purity of some food
        colours, emulsifiers, stabilizers, anticaking agents, and
        certain other food additives. FAO Nutrition Meetings
        Report Series, No. 46B, 1970; WHO/Food Add/70.37.

22.   Evaluation of food additives; specifications for the iden-
        tity and purity of food additives and their toxicological
        evaluation; some extraction solvents and certain other
        substances; and a review of the technological efficacy of
        some antimicrobial agents (Fourteenth report of the Joint
        FAO/WHO Expert Committee on Food Additives). FAO Nutrition
        Meetings Report Series, No. 48, 1971; WHO Technical Report
        Series, No. 462, 1971.

23.   Toxicological evaluation of some extraction solvents and
        certain other substances. FAO Nutrition Meetings Report
        Series, No. 48A, 1971; WHO/Food Add/70.39.

24.   Specifications for the identity and purity of some extrac-
        tion solvents and certain other substances. FAO Nutrition
        Meetings Report Series, No. 48B, 1971; WHO/Food Add/70.40.

25. A review of the technological efficacy of some antimicro-
     bial agents. FAO Nutrition Meetings Report Series, No.
     48C, 1971; WHO/Food Add/70.41.

26. Evaluation of food additives; some enzymes, modified
     starches, and certain other substances; toxicological eval-
     uations and specifications and a review of the technological
     efficacy of some antioxidants (Fifteenth report of the
     Joint FAO/WHO Expert Committee on Food Additives). FAO
     Nutrition Meetings Report Series, No. 50, 1972; WHO
     Technical Report Series, No. 488, 1972.

27. Toxicological evaluation of some enzymes, modified
     starches, and certain other substances. FAO Nutrition
     Meetings Report Series, No. 50A, 1972; WHO Food Additives
     Series, No. 1, 1972.

28. Specifications for the identity and purity of some enzymes
     and certain other substances. FAO Nutrition Meetings
     Report Series, No. 50B, 1972; WHO Food Additives Series, No.
     2, 1972.

29. A review of the technological efficacy of some antioxidants
     and synergists. FAO Nutrition Meetings Report Series, No.
     50C, 1972; WHO Food Additives Series, No. 3, 1972.

30. Evaluation of certain food additives and the contaminants
     mercury, lead, and cadmium (Sixteeth report of the Joint
     FAO/WHO Expert Committee on Food Additives). FAO Nutrition
     Meetings Report Series, No. 51, 1972; WHO Technical Report
     Series, No. 505, 1972.

31. Evaluation of mercury, lead, cadmium, and the food
     additives amaranth, diethylpyrocarbamate, and octyl
     gallate. FAO Nutrition Meetings Report Series, No. 51A,
     1972; WHO Food Additives Series, No. 4, 1972.

32. Toxicological evaluation of certain food additives with a
     review of general principles and of specifications (Seven-
     teenth report of the Joint FAO/WHO Expert Committee on Food
     Additives). FAO Nutrition Meetings Report Series, No. 53,
     1974; WHO Technical Report Series, No. 539, 1974, and
     corrigendum (out of print).

33. Toxicological evaluation of certain food additives includ-
     ing anticaking agents, antimicrobials, antioxidants, emulsi-
     fiers, and thickening agents. FAO Nutrition Meetings
     Report Series, No. 53A, 1974; WHO Food Additives Series, No.
     5, 1974.

34. Specifications for identity and purity of thickening
     agents, anticaking agents, antimicrobials, antioxidants and
     emulsifiers. FAO Food and Nutrition Paper, No. 4, 1978.

35. Evaluation of certain food additives (Eighteenth report of the Joint FAO/WHO Expert Committee on Food Additives). FAO Nutrition Meetings Report Series, No. 54, 1974; WHO Technical Report Series, No. 557, 1974, and corrigendum.

36. Toxicological evaluation of some food colours, enzymes, flavour enhancers, thickening agents, and certain other food additives. FAO Nutrition Meetings Report Series, No. 54A, 1975; WHO Food Additives Series, No. 6, 1975.

37. Specifications for the identity and purity of some food colours, flavour enhancers, thickening agents, and certain food additives. FAO Nutrition Meetings Report Series, No. 54B, 1975; WHO Food Additives Series, No. 7, 1975.

38. Evaluation of certain food additives; some food colours, thickening agents, smoke codensates, and certain other substances (Nineteenth report of the Joint FAO/WHO Expert Committee on Food Additives). FAO Nutrition Meetings Report Series, No. 55, 1975; WHO Technical Report Series, No. 576, 1975.

39. Toxicological evaluation of some food colours, thickening agents, and certain other substances. FAO Nutrition Meetings Report Series, No. 55A, 1975; WHO Food Additives Series, No. 8, 1975.

40. Specifications for the identity and purity of certain food additives. FAO Nutrition Meetings Report Series, No. 55B, 1976; WHO Food Additives Series, No. 9, 1976.

41. Evaluation of certain food additives (Twentieth report of the Joint FAO/WHO Expert Committee on Food Additives). FAO Food and Nutrition Series, No. 1, 1976; WHO Technical Report Series, No. 599, 1976.

42. Toxicological evaluation of certain food additives. WHO Food Additives Series, No. 10, 1976.

43. Specifications for the identity and purity of some food additives. FAO Food and Nutrition Series, No. 1B, 1977; WHO Food Additives Series, No. 11, 1977.

44. Evaluation of certain food additives (Twenty-first report of the Joint FAO/WHO Expert Committee on Food Additives). WHO Technical Report Series, No. 617, 1978.

45. Summary of toxicological data of certain food additives. WHO Food Additives Series, No. 12, 1977.

46. Specifications for identity and purity of some food additives, including antioxidants, food colours, thickeners, and others. FAO Nutrition Meetings Report Series, No. 57, 1977.

47.   Evaluation of certain food additives and contaminants
      (Twenty-second report of the Joint FAO/WHO Expert Committee
      on Food Additives). WHO Technical Report Series, No. 631,
      1978.

48.   Summary of toxicological data of certain food additives and
      contaminants. WHO Food Additives Series, No. 13, 1978.

49.   Specifications for the identity and purity of certain food
      additives. FAO Food and Nutrition Paper, No. 7, 1978.

50.   Evaluation of certain food additives (Twenty-third report
      of the Joint FAO/WHO Expert Committee on Food Additives).
      WHO Technical Report Series, No. 648, 1980, and corrigenda.

51.   Toxicological evaluation of certain food additives. WHO
      Food Additives Series, No. 14, 1980.

52.   Specifications for identity and purity of food colours,
      flavouring agents, and other food additives. FAO Food and
      Nutrition Paper, No. 12, 1979.

53.   Evaluation of certain food additives (Twenty-fourth report
      of the Joint FAO/WHO Expert Committee on Food Additives).
      WHO Technical Report Series, No. 653, 1980.

54.   Toxicological evaluation of certain food additives. WHO
      Food Additives Series, No. 15, 1980.

55.   Specifications for identity and purity of food additives
      (sweetening agents, emulsifying agents, and other food
      additives). FAO Food and Nutrition Paper, No. 17, 1980.

56.   Evaluation of certain food additives (Twenty-fifth report
      of the Joint FAO/WHO Expert Committee on Food Additives).
      WHO Technical Report Series, No. 669, 1981.

57.   Toxicological evaluation of certain food additives. WHO
      Food Additives Series, No. 16, 1981.

58.   Specifications for identity and purity of food additives
      (carrier solvents, emulsifiers and stabilizers, enzyme prep-
      arations, flavouring agents, food colours, sweetening
      agents, and other food additives). FAO Food and Nutrition
      Paper, No. 19, 1981.

59.   Evaluation of certain food additives and contaminants
      (Twenty-sixth report of the Joint FAO/WHO Expert Committee
      on Food Additives). WHO Technical Report Series, No. 683,
      1982.

60.   Toxicological evaluation of certain food additives. WHO
      Food Additives Series, No. 17, 1982.

61.   **Specifications for the identity and purity of certain food additives.** FAO Food and Nutrition Paper, No. 25, 1982.

62.   **Evaluation of certain food additives and contaminants** (Twenty-seventh report of the Joint FAO/WHO Expert Committee on Food Additives). WHO Technical Report Series, No. 696, 1983, and corrigenda.

63.   **Toxicological evaluation of certain food additives and contaminants.** WHO Food Additives Series, No. 18, 1983.

64.   **Specifications for the identity and purity of certain food additives.** FAO Food and Nutrition Paper, No. 28, 1983.

65.   **Guide to specifications—General notices, general methods, identification tests, test solutions, and other reference materials.** FAO Food and Nutrition Paper, No. 5, Rev. 1, 1983.

66.   **Evaluation of certain food additives and contaminants** (Twenty-eighth report of the Joint FAO/WHO Expert Committee on Food Additives). WHO Technical Report Series, No. 710, 1984.

67.   **Toxicological evaluation of certain food additives and contaminants.** WHO Food Additives Series, No. 19, 1984.

68.   **Specifications for the identity and purity of certain food additives.** FAO Food and Nutrition Paper, No. 31/1, 1984.

69.   **Specifications for the identity and purity of certain food additives.** FAO Food and Nutrition Paper, No. 31/2, 1984.

70.   **Evaluation of certain food additives and contaminants** (Twenty-ninth report of the Joint FAO/WHO Expert Committee on Food Additives). WHO Technical Report Series, No. 733, 1986.

71.   **Specifications for the identity and purity of certain food additives.** FAO Food and Nutrition Paper, No. 34, 1986.

72.   **Toxicological evaluation of certain food additives and contaminants.** WHO Food Additives Series, No. 20. Cambridge University Press, 1987.

73.   **Evaluation of certain food additives and contaminants.** (Thirtieth report of the Joint FAO/WHO Expert Committee on Food Additives). WHO Technical Report Series, No. 751, 1987.

74.   **Toxicological evaluation of certain food additives and contaminants.** WHO Food Additives Series, No. 21. Cambridge University Press, 1987.

75. Specifications for the identity and purity of certain food additives. FAO Food and Nutrition Paper, No. 37, 1987.

76. Principles for the safety assessment of food additives and contaminants in food. WHO Environmental Health Criteria, No. 70. Geneva, World Health Organization, 1987.

77. Evaluation of certain food additives and contaminants. (Thirty-first report of the Joint FAO/WHO Expert Committee on Food Additives). WHO Technical Report Series, No. 759, 1987.

ANNEX 2

ABBREVIATIONS USED IN THE MONOGRAPHS

ACTH	adenocorticotropic hormone
ADI	acceptable daily intake
BMR	basal metabolic rate
BSP	bromsulphalein
BUN	blood urea nitrogen
b.w.	body weight
CHO	Chinese hampster ovary
CNS	central nervous system
DMAB	<u>para</u>-dimethylaminoazobenzene
DMH	1,2-dimethylhydrazine
ECG	electrocardiogram
ECT	electric convulsive therapy
EEG	electroencephalogram
ERG	electroretinogram
FAO	Food and Agriculture Organization of the United Nations
FDA	Food and Drug Administration (U.S.)
FSH	follicle-stimulating hormone
GABA	gamma-aminobutyric acid
GOT	see SGOT
GPT	see SGPT
Hb	haemoglobin
HCG	human chorionic gonadotrophin
HPLC	high pressure liquid chromatography
IARC	International Agency for Research on Cancer
i.m.	intramuscular
i.p.	intraperitoneal
IPCS	International Programme on Chemical Safety
i.v.	intravenous

JECFA	Joint FAO/WHO Expert Committee on Food Additives
LD_{50}	lethal dose, median
LDH	lactate dehydrogenase
LH	leutenizing hormone
LHRF	leutenizing hormone releasing factor
LHRH	leutenizing hormone releasing hormone
LS 1	liquid smoke preparation 1
LS 2	liquid smoke preparation 2
MCA	3-methylcholanthrene
MSG	monosodium glutamate
MSH	melanocyte-stimulating hormone
NOEL	no-observed-effect level
PAHs	polynuclear aromatic hydrocarbons
PCV	haematocrit
ppm	parts per million
PTH	parathormone
PTWI	provisional tolerable weekly intake
RBC	red blood cell (erythrocyte count)
s.c.	subcutaneous
SCE	sister chromatid exchange
SG	specific gravity
SGOT	serum glutamate-oxaloacetate transaminase
SGPT	serum glutamate-pyruvate transaminase
T_3	triiodothyronine
T_4	thyroxine
TRH	TSH-releasing hormone
TSH	thyrotropin
WBC	white blood cell (total leukocyte count)
WHO	World Health Organization
w/v	weight/volume
w/w	weight/weight

ANNEX 3

JOINT FAO/WHO EXPERT COMMITTEE ON FOOD ADDITIVES
Geneva, 16-25 February 1987

Members invited by FAO

Mr J.F. Howlett, Principal Scientific Officer, Food Science Division,
Ministry of Agriculture, Fisheries and Food, London, England

Mr A.M. Humphrey, Bush Boake Allen, London, England (FAO Consultant)

Mrs D.C. Kirkpatrick, Director, Bureau of Chemical Safety, Health and
Welfare Canada, Ottawa, Canada

Professor K. Kojima, College of Environmental Health, Azubu University,
Sagamihara-shi, Japan

Dr R. Mathews, Director, Food Chemicals Codex, National Academy of
Sciences, Washington, DC, USA

Mrs I. Meyland, Scientific Officer, Central Laboratory Division A,
Nutrients and Food Additives, National Food Agency, Soborg,
Denmark

Dr J.P. Modderman, Division of Food Chemistry and Technology, Food and
Drug Administration, Department of Health and Human Services,
Washington, DC, USA (<u>Vice-Chairman</u>)

Professor F. Pellerin, Faculty of Pharmacy, Université de Paris-Sud Chatenay-Malabry, France

Members invited by WHO

Professor E.A. Bababunmi, Department of Biochemistry, College of Medicine, University of Ibadan, Ibadan, Nigeria (Rapporteur)

Dr H. Blumenthal, Director, Division of Toxicology, Center for Food Safety and Applied Nutrition, Food and Drug Administration, Washington, DC, USA

Dr B.H. MacGibbon, Senior Principal Medical Officer, Division of Toxicology and Environmental Protection, Department of Health and Social Security, London, England

Dr G. Nazario, Scientific Adviser, National Secretariat of Sanitary Surveillance, Ministry of Health, Brasilia, Brazil

Professor M.J. Rand, Professor of Pharmacology, Department of Pharmacology, University of Melbourne, Victoria, Australia (Chairman)

Dr P. Shubik, Senior Research Fellow, Green College, Oxford, England

Dr V.A. Tutelyan, Deputy Director, Institute of Nutrition, Academy of Medical Sciences of the USSR, Moscow, USSR

Secretariat

Dr J.R.P. Cabral, Scientist, International Agency for Research on Cancer, Lyons, France (WHO Temporary Adviser)

Mr A. Feberwee, Chairman, Codex Committee on Food Additives; and Deputy Director, Nutrition and Quality Affairs, Ministry of Agriculture and Fisheries, The Hague, The Netherlands (Member of FAO Secretariat)

Professor C.L. Galli, Head, Toxicology Laboratory, Institute of Pharmacology, University of Milan, Milan, Italy (WHO Temporary Adviser)

Mr R. Haigh, Principal Administrator, Commission of the European Communities, Brussels, Belgium (Temporary Adviser)

Dr Y. Hayashi, Chief, Division of Pathology, National Institute of Hygienic Sciences, Biological Safety Research Center, Setagaya-ku, Tokyo, Japan (WHO Temporary Adviser)

Dr J.L. Herrman, Division of Food and Color Additives, Center for Food Safety and Applied Nutrition, Food and Drug Administration, Washington, DC, USA (WHO Consultant)

Dr L.G. Ladomery, Food Standards Officer, FAO/WHO Food Standards Programme, Food Policy and Nutrition Division, FAO, Rome, Italy

Dr M. Mercier, Manager, International Programme on Chemical Safety, Division of Environmental Health, WHO, Geneva, Switzerland

Dr A.W. Randell, Nutrition Officer (Food Science), Food Policy and Nutrition Division, FAO, Rome, Italy (Joint Secretary)

Dr S.I. Shibko, Associate Director for Toxicological Evaluation, Division of Toxicology, Center for Food Safety and Applied Nutrition, Food and Drug Administration, Washington, DC, USA (WHO Temporary Adviser)

Dr G. Vettorazzi, Senior Toxicologist, International Programme on Chemical Safety, Division of Environmental Health, WHO, Geneva, Switzerland (Joint Secretary)

Professor R. Walker, Professor of Food Science, Department of Biochemistry, University of Surrey, Guildford, England (WHO Temporary Adviser)

ACCEPTABLE DAILY INTAKES, OTHER TOXICOLOGICAL
RECOMMENDATIONS, AND INFORMATION ON SPECIFICATIONS

Substance	Specifications [1]	ADI for man and other toxicological recommendations
A. Food additives		
Enzyme preparations		
alpha-Amylase from A. oryzae	N,T	Acceptable [2]
Protease from A. oryzae	R,T	Acceptable [2]
Amyloglucosidases from A. niger	N,T	0-1 mg/kg b.w. [3]
beta-Glucanase from A. niger	N,T	0-1 mg/kg b.w. [3]
hemi-Cellulase from A. niger	N,T	0-1 mg/kg b.w. [3]
Pectinases from A. niger	N,T	0-1 mg/kg b.w. [3]
Protease from A. niger	O	0-1 mg/kg b.w. [3]
beta-Glucanase from T. harzianum	N,T	0-0.5 mg/kg b.w. [3,4]
Cellulase from T. reesei	N,T	0-0.3 mg/kg b.w. [3,4]
Cellulase from P. funiculosum	N,T	No ADI allocated [5]
Pectinase from A. alliaceus	O	No ADI allocated [5]
Flavouring agents		
trans-Anethole	S	0-2.5 mg/kg b.w. [4]
Benzyl acetate	S	0-5 mg/kg b.w. [4]
Smoke flavourings	N,T	Provisional acceptance [6]
Food colours		
Beet red	R	ADI "not specified" [7]
Canthaxanthin	R	0-0.05 mg/kg b.w. [4]
Carbon black	R,T	No ADI allocated [5,8]
Carotenes (algae)	N,T [9]	No ADI allocated [5]
Carotenes (vegetable)	N,T [9]	No ADI allocated [5]
Citranaxanthin	R	No ADI allocated [5]
Xanthophylls (mixed carotenoids)	N,T [10]	No ADI allocated [5]
Xanthophylls (Tagetes extract)	N,T [10]	No ADI allocated [5]

Substance	Specifications[1]	ADI for man and other toxicological recommendations

Miscellaneous food additives

Ferrous gluconate	S	0.8 mg/kg b.w.[11]
Glutamic acid and its salts	R	ADI "not specified"[7,12]
4-Hydroxymethyl-2,6-ditert-butylphenol	W	No ADI allocated[5]
Polydextroses	S	ADI "not specified"[7]
Tannic acid	R,T	ADI "not specified"[4,7,13]

B. **Contaminants**

Aflatoxins	–	Lowest practicable level[14]

Specifications only[1]

Activated carbon	R
alpha-Amylase and glucoamylase from A. oryzae	R,T
beta-Carotene, synthetic	R
Brilliant black BN	S
Caramel colours	R
Carthamus yellow	R
Chlorophylls	R
Chlorophylls, copper complexes	R
Chlorophyllins, copper complexes, sodium and potassium salts	R
Hydrogenated glucose syrups	R,T[15]
Insoluble polyvinylpyrrolidone	S
Paprika oleoresin	R
Patent blue V	R
Polyethylene glycols	R
Polyglycerol esters of fatty acids	S
Ponceau 4R	S
Potassium bromate	R
Potassium dihydrogen citrate	R
Riboflavin	R
Riboflavin 5'-phosphate, sodium	R
Sucrose esters of fatty acids	S
Talc	S
Triammonium citrate	S
Xylitol	R,T

Notes to Annex 4

1.　N, new specifications prepared; O, specifications not prepared; R, existing specifications revised; S, specifications exist, revision not considered or not required; T, the existing, new or revised specifications are tentative and comments are invited; and W, previously established specifications withdrawn.

2.　Acceptable for use in food processing. These enzymes are derived from microorganisms that are traditionally accepted as constituents of foods or are normally used in the preparation of foods. These products are regarded as foods and, consequently, considered acceptable, provided that satisfactory chemical and microbiological specifications can be established.

3.　Based on the percentage of T.O.S. (total organic solids); % T.O.S. = 100 - (A + W + D), where A = % ash, W = % water, and D = % diluent and carrier.

4.　Temporary acceptance.

5.　Insufficient information available on its toxicology and/or chemical composition to establish an ADI.

6.　Analytical and compositional data, including data on variability, are required; further safety studies on a well-defined spectrum of smoke flavourings are desired. Benzo(a)pyrene should not exceed 10 μg per kg.

7.　ADI "not specified" means that, on the basis of the available data (chemical, biochemical, toxicological, and other), the total daily intake of the substance, arising from its use at the levels necessary to achieve the desired effect and from its acceptable background in food, does not, in the opinion of the Committee, represent a hazard to health. For that reason, and for the reasons stated in the individual evaluations, the establishment of an ADI expressed in numerical form is not deemed necessary.

8.　The use of carbon black from hydrocarbon sources in food contact materials is provisionally accepted.

9.　The previous specifications for carotenes (natural) were withdrawn.

10.　The previous specifications for xanthophylls were withdrawn.

11.　Provisional maximum tolerable daily intake (PMTDI) for iron.

12.　Group ADI for L-glutamic acid and its ammonium, calcium, magnesium, monosodium, and potassium salts.

13.   For use as a processing aid.

14.   Presence in food should be reduced to irreducible levels.
An irreducible level is defined as that concentration of a
substance that cannot be eliminated from a food without
involving the discarding of that food altogether, severely
compromising the ultimate availability of major food
supplies.

15.   This specification also covers dry, food grade maltitol,
which is the substance listed on the agenda of this meeting.

12 301

For EU product safety concerns, contact us at Calle de José Abascal, 56–1°,
28003 Madrid, Spain or eugpsr@cambridge.org.

www.ingramcontent.com/pod-product-compliance
Ingram Content Group UK Ltd.
Pitfield, Milton Keynes, MK11 3LW, UK
UKHW012315141225
465965UK00001B/75